# A THEORY OF PHYSICAL PROBABILITY

RICHARD JOHNS

# A Theory of Physical Probability



UNIVERSITY OF TORONTO PRESS
Toronto Buffalo London

Toronto Buffalo London

ISBN 0-8020-3603-1

Printed on acid-free paper



Toronto Studies in Philosophy
Editors: James R. Brown and Amy Mullin

---

**National Library of Canada Cataloguing in Publication Data**

Johns, Richard, 1968–
A theory of physical probability

(Toronto studies in philosophy)
Includes bibliographical references and index.
ISBN 0-8020-3603-1

1. Chance. I. Title. II. Series.

BC141.J64 2002 123′.3 C2001-903934-4

---

This book has been published with the help of a grant from the Humanities and Social Sciences Federation of Canada, using funds provided by the Social Sciences and Humanities Research Council of Canada.

University of Toronto Press acknowledges the financial assistance to its publishing program of the Canada Council for the Arts and the Ontario Arts Council.

University of Toronto Press acknowledges the financial support for its publishing activities of the Government of Canada through the Book Publishing Industry Development Program (BPIDP).

# Contents

1 Introduction 3
1.1 Motivation for the Causal Theory of Chance 3
1.2 Problems for the Causal Theory 5
1.3 Advantages of the Causal Theory 5
1.4 Overview of the Book 6

2 Logic and Probability 9
2.1 The Objections to Logical Probability 10
2.2 The Nature of Logical Probability 13
2.3 Measuring Degrees of Belief 29
2.4 The Axioms of Probability 33
2.5 Relative Probabilities 36
2.6 Interval Probabilities 37
2.7 The Symmetry Axiom 38

3 Causation and Determination 53
3.1 Causation 53
3.2 Determination 64
3.3 How Are Causation and Determination Related? 70

4 Physical Chance 84
4.1 The Definition of Chance 84
4.2 Chance Is Relativized to a System 87
4.3 Lewis's Objections 88
4.4 A Proof of Miller's Principle 89
4.5 The Objections of Howson and Urbach 90
4.6 Chance and Relative Frequency 91

4.7 Frequency Theories of Probability 96
4.8 Conditional Chances 106

5 Classical Stochastic Mechanics 109
5.1 What Is CSM Good for? 109
5.2 The Law Function 110
5.3 Relevance and Correlation 114
5.4 Chance in a Composite System 118
5.5 Sub-histories, States, and Markov Systems 121
5.6 Boundary Conditions and Time 123
5.7 The Arrow of Time 133

6 Correlation 148
6.1 Classical and Quantum Correlation 149
6.2 Reactions to EPR 167
6.3 Beyond Postulate CSM3 181

7 The State Vector 188
7.1 The Problem 188
7.2 Large and Small Systems 206
7.3 Chance for Small Systems 220
7.4 Summary 230

8 Conclusion 233

NOTES 235
BIBLIOGRAPHY 245
INDEX 253

# A THEORY OF PHYSICAL PROBABILITY

# 1

# Introduction

## 1.1 MOTIVATION FOR THE CAUSAL THEORY OF CHANCE

There is, at present, no satisfactory theory of physical probability, or chance. In this book I offer such a theory, called the causal theory of chance (or CTC for short). The term *chance* has had a number of different meanings, but is now regularly used to refer to the kind of probability that is inherent in some physical systems. The half-life of a nucleus, for instance, is defined as the length of time in which the probability of decay is $\frac{1}{2}$. The use of the term 'probability' here has nothing to do with any weight of evidence or degree of subjective certainty, as is the case in some other contexts, but appears to denote something objective and physical. Thus, when we say that in the period of its half-life the chance of a nucleus decaying is $\frac{1}{2}$, we mean this in an objective, physical sense.

It is generally held that, if an event E is physically determined to occur, then its chance is one. (Also, if it is determined not to occur, then its chance is zero.) But when we say that E is determined, what do we mean that it is determined *by*? We are saying, of course, that E is determined, or nomically necessitated, by the things that *cause* E. An event has a physical probability of one just in case its occurrence is nomically necessary, given some prior cause.

If this is true, then a *chancy* event (one which has a chance that differs from zero and one) is not determined to occur by any prior cause. There seem to be two quite different ways to accommodate such chancy events. First, one could maintain that they are uncaused, or at least not fully caused. A partially determined event would be one that is only partially caused.[1] This view allows one to hold onto the common opinion that a total cause always determines its effect. From this point of view, chancy events may be described as 'accidental,' 'capricious,' 'glitchy,' or 'mere hap.'[2] I prefer to use the expression 'events from nowhere,' however, as these events would have no ontological source. They would not arise from other events; they would not be produced, or generated, by anything else, but would simply come into existence all by themselves. The second possible view about chancy events is that they all have total causes, just like determined events, but they are not determined to occur by those causes.

In this book I develop a version of the second view about chancy events, that such events are fully caused, but not determined by those causes. Why do I take this route, rather than its alternative? The main reason is that I consider the notion of an uncaused event to be highly implausible.[3] Things in spacetime do not simply happen – they are always caused to occur by something prior. A second reason for holding chancy events to be fully caused is that their chances are often (although perhaps not always) capable of being identified with fairly precise numerical values. Even if the occurrence of an event E is not determined, the chance of E may nonetheless be fully determined, at least to a high level of accuracy.[4] Chances, at least in simple physical experiments, often seem to be quite stable and regular, and not at all accidental or capricious. If a chancy event has no cause, then where does this precise value come from? What is its source? A third reason is more technical, and cannot be substantiated until chapter 6. I will argue that the observed correlations between some pairs of causally isolated chancy events, such as in the Aspect experiments,[5] require that each separate event be fully caused.

If we take the view that a chancy event is one that is not necessitated by its causes, then what theory of chance does this lead to? Since chances form a range of values between zero and one, we require there to be a corresponding range of *degrees* of determination, or nomic necessitation. The extremes of this range are the cases where E is determined to occur, and where E is determined not to occur (by its causes). In intermediate cases, where E has some chance $p$ of occurring, E is determined to occur to the degree $p$. In short, my claim is that the chance of an event is the degree to which it is determined by its cause.

The idea that causes need not determine their effects is strongly associated with a certain view of causation, known as the probabilistic theory of causation. One might therefore think that I would adhere to some version of the probabilistic theory of causation, but this is not the case. The aim of such theories is to give an analysis of causation, taking chance to be primitive. In contrast, I take causation as primitive and try to give an analysis of chance. Moreover, this direction of analysis is not a matter of choice, as it is with concepts that are inter-definable, such as possibility and necessity. Chance can be analysed using causation, whereas causation cannot be analysed (correctly) in terms of chance. I hold that the probabilistic theories of causation are all false.

It is in order to emphasize the contrast between my theory of chance and probabilistic theories of causation that I call my theory the causal theory of chance. It is, if you will, the exact reverse of a chancy theory of causation.

## 1.2 PROBLEMS FOR THE CAUSAL THEORY

The main difficulty faced by the causal theory of chance is that its acceptance requires some significant innovations in metaphysics and logic. The metaphysical problem, dealt with in chapter 3, is that the causal theory requires a very sharp distinction between the relations of causation and determination. The logical problem, discussed in chapter 2, is that the causal theory requires there to be a relation of partial logical entailment, or logical probability.

These problems are easily seen, even from the simple presentation of CTC given in the previous section, that the chance of an event is the degree to which it is determined by its causes. The need for logical probability arises from the appeal in CTC to degrees of determination. According to several standard accounts of the determination relation,[6] which I follow, determination is simply logical entailment relative to the dynamical laws. Thus, a degree of determination is a degree of partial entailment, or logical probability, relative to those same laws.

CTC also clearly requires that even a total cause need not determine its effects. As will be shown by the symmetry-conservation argument of §3.3.4, it is very hard to avoid the common position that an undetermined event must also be uncaused. Indeed, to escape from the conclusion of this argument it is necessary to develop a new distinction between completeness and maximality, regarding possible states of affairs. This distinction is, however, very useful in the remainder of this book, since I argue (for instance) that the quantum-mechanical state is maximal although incomplete.

## 1.3 ADVANTAGES OF THE CAUSAL THEORY

I claimed above that there is, at present, no satisfactory theory of chance. What I meant by this is that there is no theory that accounts for the properties of chance that are generally recognized, which I take to be as follows:

1 Chance supervenes on the physical properties of the system in question.
2 Chance is a probability function, that is, a normalized measure.
3 Chance is defined for single events.

4 It makes sense to condition the chance function upon an event.
5 Knowledge of the chance of an event *A* authorizes a numerically equal degree of belief that *A*; that is, Miller's Principle holds.
6 Chances can be discovered, approximately and fallibly, by measuring relative frequencies.
7 The probabilities in quantum mechanics are chances (at least where pure states are concerned).

As is argued in chapter 4, frequency theories of physical probability cannot account for properties 5 and 6, so that it is now common to take chance as a primitive quantity and Miller's Principle as a fundamental axiom. Within my theory, by contrast, Miller's Principle is proved as a theorem (see Theorem 4.4.1), and properties 1, 2, 3, 4, and 6 are also shown to hold. Moreover, as argued in chapters 6 and 7, CTC enables one to make sense of the behaviour of chances in quantum theory.

The causal theory of chance, therefore, accounts for all the generally accepted properties of physical chance. This is no mean achievement, since these constraints are very difficult to satisfy simultaneously. In particular, it is hard to reconcile the physical and epistemological aspects of chance, as represented by properties 1 and 5.

## 1.4 OVERVIEW OF THE BOOK

The remainder of this book divides into three main sections. First, in chapters 2 and 3, some preliminary problems are addressed in order to prepare the ground for CTC. The causal theory of chance is then stated and defended in chapter 4. Finally, in chapters 5 to 7, CTC is applied to some outstanding problems in theoretical physics.

The causal theory of chance requires certain innovations in logic, principally the introduction of a relation of partial entailment, or logical probability. The consensus in many quarters today is that there is no probability function that is determined by logic alone, so the first task is to show that this view is mistaken. In chapter 2 I attempt to demonstrate that logical probability is still very much a viable idea by giving an account of it. In order to develop this account, a new philosophy of logic is needed, which is based on a suggestion of Peter Gärdenfors (1988: 135). This system of logic, based upon epistemic states, also allows other logical devices to be introduced, such as relative propositions and logical independence. These are extremely useful throughout the book.

In chapter 3 I investigate the relations of causation and determination, a clear understanding of which is needed for CTC. In particular, I argue that causation is a relation of 'ontological support' of one concrete event by another, while determination is a logical relation between possible states of affairs. In order to understand the fact that determination is weaker than logical entailment, the notion of the generalized lagrangian is introduced, based on Aristotle's doctrine of dynamical natures. A large part of the chapter is devoted to the claim that a total cause must determine its effects. Arguments for this view are considered and refuted.

In chapter 4 the causal theory of chance is outlined in detail, and compared to its rivals. It will be shown that the causal theory entails Miller's Principle, that knowledge of the chance of an event E warrants a numerically equal degree of belief that E occurs. From this result it follows that, according to the causal theory, chances can be inferred (albeit approximately and fallibly) from empirical observations of relative frequency. This conforms perfectly to what we ordinarily suppose about chance and frequency.

The remaining three chapters concern the application of the causal theory to some outstanding problems in theoretical physics. The purpose of these chapters is twofold. First they provide support for the causal theory itself, since they demonstrate its fruitfulness in solving problems. They give indirect empirical evidence that stochastic events are caused but not determined. Second, since the problems considered are themselves important, the solutions provided are of independent interest.

In chapter 5 I set up a mechanical formalism for stochastic processes. The main purpose of this is to act as a foundation for the later chapters on correlation and measurement, but it is also used to tackle the problem of the direction of time, which I see primarily as one of accounting for temporally asymmetric physical phenomena. In particular, I focus upon Reichenbach's common cause principle as a general example of such a phenomenon. The formalism is based upon five assumptions about physical systems that seem reasonable, at least at large scales, and from these the required phenomena may be derived.

In chapters 6 and 7 I analyse some problems in quantum mechanics from the point of view of the causal theory. This involves a detailed examination of the *dual* nature of chance. According to the causal theory, the chance function supervenes on the physical properties of the system concerned, and thus is itself a *physical* feature of the system. Since, however, this function is defined using logical probability, it also turns out to be an *epistemic* probability function. These are the two faces of chance; it is entirely physical, yet it also represents rational degrees of belief within a particular possible epistemic state. A second feature of the causal theory, which also assumes central importance in understanding quantum chances, is that the knowledge of the system 'possessed' by the chance function is incomplete, even though it is maximal.

In chapter 6 I develop a theory of correlation that explains the counter-intuitive results obtained in EPR-type experiments. I show how these correlations, which cannot be used to transmit information, do not require any causal link between the separated systems. In other words, there is no need to drop Einstein's locality principle of special relativity, that causal influences cannot travel faster than light. We are forced to drop another locality principle, however, that a *maximal state* is always local, in the sense that the maximal state of a pair of systems can always be split into two separate states, one for each system.

In chapter 7 the theory of quantum correlation developed in chapter 6 is applied to the more difficult problem of interpreting the state vector, and particularly the way it changes when a measurement occurs. The state vector is closely related to the chance function, and shares the latter's dual nature, being both physical and epistemic.

# 2

# Logic and Probability

The main aim of this chapter is to argue that there is such a thing as logical probability. The conception of logical probability I defend is somewhat different from previous accounts, however, so why do I call it 'logical probability'? The essential notion of logical probability, as pioneered by Keynes (1921), is captured by the following three ideas:

1 As the name suggests, logical probabilities are probabilities that are, in some sense, part of logic.
2 A logical probability is a degree of entailment of one proposition by another. Thus, a logical probability represents a logical relation between two propositions.
3 A logical probability is some sort of degree of belief. The rough idea is that the logical probability of $A$ given $B$, which we write as $Pr(A \mid B)$, has something to do with the degree to which $A$ should be believed, for someone who knows only that $B$ holds.

Since my account of logical probability agrees with that of Keynes on these points, the use of the term 'logical probability' seems appropriate.

These three strands are far from independent, of course, as (2) fleshes out what is meant by (1), and (3) does the same for (2). Since entailment is a logical relation par excellence, it follows that degrees of entailment must be part of logic as well. What, however, is meant by a 'degree of entailment'? We obviously require an account of partial entailment according to which ordinary entailment is a special case of partial entailment. The idea of a degree of entailment as a conditional degree of belief seems, at least, to be capable of providing such an account. If $B$ entails $A$ fully, then an ideal thinker who knows $B$ with certainty will also believe $A$ to the greatest possible degree.

Thus, we see that the notion of logical probability is most fully captured by (3).

I realize that some may find the following account of logical probability unconvincing, and so it may be wondered which features of logical probability are required for the rest of the book. Apart from the obvious requirements, such as satisfaction of the axioms of probability, these are as follows.

(i) Logical probability is not anthropocentric (so that chance is not anthropocentric). It is defined on objective states of affairs rather than human thoughts.
(ii) The Authority of Logic principle of §2.2.3 holds.

These two properties of logical probability are needed to ground the two main features of physical chance. (i) is needed for chance to be an objective, physical quantity, and thus not dependent upon human beings, and (ii) is required for Miller's principle, that is, the Authority of Chance principle, that knowledge of the chance authorizes a numerically equal degree of belief. If another account of logical probability were found that also satisfied (i) and (ii), then it would serve the purposes of this book equally well.

## 2.1 THE OBJECTIONS TO LOGICAL PROBABILITY

Logical probability has been in the academic wilderness for some decades now, and many consider that its rightful place. Before I begin to investigate it, and examine its properties, I therefore should explain why I think there is at least some hope of giving a satisfactory account of it, despite the failure of greater minds to do so. Perhaps the main reason why the prospects for logical probability seem so dim is that previous attempts to characterize it have failed, and in ways that seem quite fatal. The principal difficulties are as follows.

I Measures that are arguably logical in character do not support learning from experience, and
II Logical probability seems to require an indifference principle, but these all lead to contradictions.

In addition to these practical problems that arise when trying to construct a theory of logical probability, there are also two more theoretical objections to the concept itself, namely:

III The logical probability function is too opinionated to be part of logic.

The number $Pr(A \mid T)$, where $T$ is a tautology, is an exact degree of belief in the possible state of affairs $A$ on the basis of no information at all! Logic, however, is independent of matters of fact, and so should have no such opinions.

IV The idea of logical probability violates the distinction between logic and psychology. A probability is a degree of belief, and belief belongs to psychology rather than logic. Logic is concerned with truth, not belief, and there are no degrees of truth.

These problems, as a whole, may look quite severe, but they will all be answered in this chapter. Objection II is met by formulating a symmetry axiom, proving it from the definition of logical probability, and showing that it is free of contradiction. The symmetry relation involved, which exists between some pairs of states of affairs, certainly appears to be logical in character.

Objections I and IV are the most interesting, as they concern the relation of logic to human thought. Problem I assumes, for instance, that logical probability is involved in the rationality of scientific inferences. This assumption was certainly made by the two main investigators of logical probability, J.M. Keynes (1921) and R. Carnap (1950). Keynes, for instance, summarizes the notion of logical probability as follows:

> Part of our knowledge we obtain direct; and part by argument. The Theory of Probability is concerned with that part which we obtain by argument, and it treats of the different degrees in which the results so obtained are conclusive or inconclusive ... Given the body of direct knowledge which constitutes our ultimate premisses, this theory tells us what further rational beliefs, certain or probable, can be derived by valid argument from our direct knowledge. This involves purely logical relations between the propositions which embody our direct knowledge and the propositions about which we seek indirect knowledge. (3, 4)

Keynes believed, in other words, that the inductive inferences of natural science can be analysed using logical probabilities. He is quite explicit about this, as his main aim in the *Treatise* is to lay a firm, logical foundation for scientific reasoning. In science one often claims that a theory $H$ is supported by experimental data $E$, that it is valid to accept the theory in the light of the empirical evidence. In Keynes's view this simply means $E$ entails $H$ to some degree, so that $Pr(H \mid E)$ is reasonably high.[1]

This view of Keynes, that warranted scientific theories have high logical probabilities given the empirical data, is now known to be false. The basic problem is neatly illustrated by the example of predicting the colours of balls

chological relations. The subject matter of logic is the human mind, and the particular way it forms concepts, makes judgments, performs inferences, and so on. Frege is implacably opposed to this invasion of psychological terms into logic and the foundations of mathematics, so that the first several pages of the *Grundlagen* are devoted to a forthright, sarcastic denunciation of it. He writes, for instance:

> A proposition may be thought, and again it may be true; let us never confuse these two things. We must remind ourselves, it seems, that a proposition no more ceases to be true when I cease to think of it than the sun ceases to exist when I shut my eyes. Otherwise, in proving the Pythagorean theorem we should be reduced to allowing for the phosphorus content of the human brain; and astronomers would hesitate to draw any conclusions about the distant past, for fear of being charged with anachronism, – with reckoning twice two as four regardless of the fact that our idea of number is a product of evolution and has a history behind it. It might be doubted whether by that time it had progressed so far. How could they profess to know that the proposition $2 \times 2 = 4$ already held good in that remote epoch? Might not the creatures then extant have held the proposition $2 \times 2 = 5$, from which the proposition $2 \times 2 = 4$ was only evolved later through a process of natural selection in the struggle for existence? Why, it might even be that $2 \times 2 = 4$ is itself destined in the same way to develop into $2 \times 2 = 3$! *Est modus in rebus, sunt certi denique fines!* (1884: v–vi)

The distinction between logic and psychology, that is, truth and belief, is of the greatest importance, in my view, and an excellent place to begin an enquiry into the nature of logic. It is well known that a sentence[4] has two properties: it expresses a thought and it has a truth value. A sentence is true just in case it corresponds to an actual state of affairs in the concrete world, and is false if it corresponds to some state of affairs that does not obtain.[5] Thus, every sentence, if it is meaningful, corresponds to some state of affairs, which may or may not be actual. Indeed, we may say that the state of affairs described by a sentence is the meaning of the sentence.[6]

In human terms a sentence is used for communication, to express a belief, or rather the content of a belief. Following Frege, I shall call the content of a possible belief a *Gedanke*. Since two sentences may express the same belief, or have the same content, it is possible for two sentences to share the same Gedanke. If two Gedanken are equal, then it is not possible for a (properly functioning) human to believe one and not the other, at a single time. Like Frege, I assume that two humans may think the same (type of) Gedanke, and that to understand a sentence is to grasp the Gedanke it expresses.

One might be tempted to say that a Gedanke, or belief content, is the thing believed when one has that belief. This would be a grave mistake, however. To explain this point I shall adapt one of Frege's illustrations (1892: 60). Suppose one looks at the moon through a refracting telescope. The objective lens of this instrument forms a virtual image of the moon just in front of the eye lens. When one looks into the telescope, it is the moon that one observes, the actual celestial body. The image inside the telescope is, however, a necessary part of the physical process of observation; one might say that the moon is presented to the observer via this virtual image. In a similar way, consider a belief that the moon has no atmosphere. This belief concerns the external world, that is, the moon itself, and not any psychological entities such as thoughts. The thing that is believed here is the possible objective state of affairs that the moon has no atmosphere. A belief, however, is a psychological event that occurs within a human mind, and a Gedanke is one necessary component of any belief. It is analogous to the virtual image within the telescope, which is a necessary part of the physics of the process of observing with a telescope. A Gedanke is the manner in which a state of affairs is presented to a human mind.

We see then that a sentence both expresses a thought, or Gedanke, and also corresponds to a state of affairs. The Gedanke is a psychological entity, a feature of the human mind, whereas a state of affairs is objective, or 'inhuman,' that is, a feature of the external world. The logic/psychology distinction therefore matches the state of affairs / Gedanke distinction. States of affairs are part of the realm of truth, or objective reality, whereas Gedanken belong to the realm of human belief. Why is it necessary to make this distinction, however? Why not identify these two entities (as they perhaps look rather similar) and simplify the picture? (One might call the single entity a 'proposition.') The necessity of this distinction is proved by Frege, however, in his monumental paper 'Über Sinn und Bedeutung' (1892b), with an argument that I will now summarize.

Frege begins by noting that equality, or identity, gives rise to questions that are not easy to answer. He observes that an identity statement '$a = b$' must refer to the subject matter, that is, the objects denoted by '$a$' and '$b$,' rather than the written symbols, or it would not express any proper knowledge of the world. In that case, however, an identity statement seems to be rather trivial, as it merely says that an object is identical to itself. Yet some identities are far from trivial, or obvious, being important discoveries. For instance, Phosphorus (the morning star) is equal to Hesperus (the evening star). This was an important astronomical discovery, requiring empirical observations, and cannot be known by internal reflection alone. How can this be?

This identity is far from trivial, Frege argues, because the two names 'Hesperus' and 'Phosphorus' are associated with different manners of presentation of the single planet Venus. Venus presents itself to us humans under two different guises, sometimes as a morning star and sometimes as an evening star, so that it is not easy to tell that it is the same planet in each case. The manner of presentation associated with a linguistic unit, such as a proper name, Frege calls a *Sinn*, whereas the real meaning of that unit is its *Bedeutung*. Thus, in the Venus example, the names 'Hesperus' and 'Phosphorus' have the same Bedeutung (the planet Venus) but distinct Sinne. Note that, while a Bedeutung is an external, inhuman object, a Sinn is some sort of psychological entity, being bound up with the manner in which an object presents itself to humans.

Now let us consider the two sentences:

1 Hesperus is Hesperus.
2 Hesperus is Phosphorus.

These two sentences have the same meaning, in that they correspond to the same state of affairs. (One often says that they are 'true in the same class of possible worlds.') They do not express the same Gedanke, however, as someone might, without any mental dysfunction, believe (1) but not (2). At the level of Gedanken, (1) is a trivial tautology, whereas (2) is a significant statement of fact. This point is illustrated by considering the following two sentences:

3 Abraham believes that Hesperus is Hesperus.
4 Abraham believes that Hesperus is Phosphorus.

It may well be that (3) is true and (4) is false. It is clear that, for Abraham at least, the two thoughts are quite distinct, even though they correspond to the same external state of affairs. If the two Gedanken were equal, then it would be possible to substitute one for the other, even in a belief context, without any change of truth value in the whole sentence.

The distinction between Gedanke and state of affairs seems to be an instance of Frege's general distinction between Sinn and Bedeutung. The Bedeutung of a linguistic unit is its real meaning, its significance in the external world. The Sinn of a linguistic unit contains the manner in which the Bedeutung of the unit is presented to humans, under that unit. Thus, the Bedeutung of 'Hesperus' is the planet itself, while the Sinn of 'Hesperus' is determined by the manner in which Venus presents itself to us under the name 'Hesperus,' that is, as a bright light visible above the western horizon after sunset. In a similar way, the Bedeutung of a sentence seems to be the state of affairs it represents, while the Sinn is the manner in which that state of affairs is presented to us humans, as a Gedanke. The sentences (1) and (2) present the same state of affairs under two different guises.

This identification of the Bedeutung of a sentence with the state of affairs it represents is Fregean in spirit, but not in detail. In what I consider to be his greatest mistake, Frege argues that the Bedeutung of a sentence is its truth value. What is his argument for this? Frege's premiss is that the Bedeutung of a sentence should not depend upon the Sinne of its sub-units (such as proper names) but only on their Bedeutungen. This premiss is surely sound, as Bedeutungen are all external objects, having nothing to do with human beings. Then consider the sentences:

5 Hesperus has no moon.
6 Phosphorus has no moon.

Whatever the Bedeutung of (5) may be, it must be equal to that of (6), since (6) is obtained from (5) by substituting 'Phosphorus' for 'Hesperus,' whose Bedeutungen are equal. Now Frege notes that the truth values of (5) and (6) are equal, and indeed that truth values must always be preserved under such a substitution. He then asks, rhetorically, 'What feature except the truth-value can be found that belongs to such sentences quite generally and remains unchanged by substitutions of the kind just mentioned?' (1892b: 64–5). For some reason that is unclear to me, Frege does not seem to notice that the state of affairs represented by a sentence is just such a feature. The sentences (5) and (6), for example, correspond to the same state of affairs, as they are true in exactly the same class of worlds.

Frege's argument thus fails to rule out states of affairs as the Bedeutungen of sentences. Is there any reason to hold that the meaning of a sentence is a state of affairs, rather than a truth value, however? The following three arguments prove this. First, consider the following true statement:

7 The 1998 World Cup was held in France.

Since (5) and (6) are also true it follows that, according to Frege, (5), (6), and (7) all have the same meaning, namely, The True. It nonetheless seems that there is a similarity of meaning that exists between (5) and (6) that does not exist between (5) and (7). One would describe this similarity by saying that, while (5) and (6) correspond to the same state of affairs, (5) and (7) represent distinct states of affairs. Frege apparently has no way to mark this important distinction, however. He cannot appeal to Sinn here, as the Sinne of these three sentences are all distinct.

A second argument proceeds from the premiss that the sentence is the basic linguistic unit, and the fundamental carrier of meaning, since it is the smallest unit that 'says something.' In that case, the meanings of smaller units, such as proper names and predicates, should be definable in terms of the meaning of a sentence. This is just not true if the meaning of a sentence is its truth value – the planet Venus does not seem to be a component of The True, for example. Venus does look as if it is a component of the state of affairs that Venus has no moon, on the other hand. It might even be possible to define Venus as the common component of all the states of affairs concerning Venus. This would explain why there is no possible state of affairs in which Venus is not identical to Venus – it is because Venus, as an object, is constituted by being a common component of distinct states of affairs. In a similar way, a Begriff (or concept, i.e., the meaning of a predicate) may perhaps be abstracted from an equivalence class of states of affairs.

The third argument is that logic should only require Bedeutungen. The primary logical relations, namely, consistency and entailment, are not relations between truth values. Does the True entail the True? Is the False consistent with the False? These questions are meaningless. These relations do hold between states of affairs, however. One state of affairs can entail, or include, another. For example, the state of affairs of Phosphorus's having no moon includes the state of Hesperus's having fewer than two moons. To make these relations of consistency and entailment part of logic, Frege is forced to regard the Sinne as part of logic rather than psychology. This is perhaps the reason why he is so adamant that Sinne are objective rather than subjective, and sharply to be distinguished from mental ideas. In view of Frege's account of a Sinn as a manner of presentation (to humans, presumably), this placement of Sinne on the logic side of the logic/psychology (truth/belief) division is rather implausible. In particular a Gedanke, or thought, is clearly a human thought, as who else's thought could it be? Thus, Sinne are closely tied to humans, and hence to psychology.

### *2.2.2 Entailment and Probability*

It is stated in the introduction that a logical probability is a degree of entailment, so we must try to understand the entailment relation. It is generally thought that entailment is a relation between propositions, but what are propositions? Sometimes 'proposition' means a declarative sentence, but sentences themselves cannot directly entail other sentences. Rather, the proposition expressed by one sentence may entail that expressed by another sentence. In the previous section we saw the need to distinguish between two different kinds of sentence meaning, namely, Gedanken and states of affairs. Perhaps either Gedanken or states of affairs are propositions.

It seems to me that the term 'proposition' is sometimes used to mean a Gedanke, and sometimes a state of affairs, without a clear distinction being made between these two. Indeed, if we wish to use the term 'proposition,' we should perhaps talk of 'logical propositions' (objective states of affairs) and 'psychological propositions' (human belief contents). There are plenty of examples of the term 'proposition' being used in each of these two ways. A proposition is often said to be defined by a class of possible worlds, namely, the set of worlds in which it holds true. In this case, a proposition is indistinguishable from a state of affairs. At other times, however, a proposition is said to be the content of a (presumably human) belief.[7]

If there are two kinds of proposition, the logical and the psychological, then we might expect there to be two matching kinds of entailment, which seems to be the case. Consider, for instance, the sentences (5) and (6) above. Does (5) entail (6)? At the level of Bedeutung they have the same meaning, so that each trivially entails the other. The state of affairs of Phosphorus's having no moon is necessitated by Hesperus's having no moon. At the level of Sinn, however, it seems that neither entails the other. It would certainly be invalid for a human to infer (6) from (5) since, as far as one knows, (5) could well be true and (6) false. We should not imagine, of course, that these two levels of entailment are quite disconnected. After all, a Gedanke is just a manner in which a human mind grasps a state of affairs.

Entailment at the level of Gedanken is a rather messy affair, as one has to take account of human limitations, including the following. First, a single state of affairs may have different probabilities (degrees of rational belief) under different Gedanken, within a single epistemic state. Second, there is no guarantee that Gedanken defined by logical operations will exist. There may be two Gedanken $A$ and $B$ that are consistent, and yet no conjunction $A$ & $B$ exists simply because it is too complicated. The human mind is finite, and cannot grasp states of affairs of arbitrary complexity, so that the class of Gedanken will not form even a Boolean algebra. Third, the same finite nature of the mind surely precludes deductive closure of epistemic states, even as an ideal.

Entailment at the level of Bedeutung, by contrast, is much cleaner and 'logical,' being free of human limitations. The class of states of affairs does (it is assumed) form a Boolean algebra, for example, as for any two consistent states of affairs there exists also a conjunction of them. If logical probability is to be used in an analysis of physical chance, it must be non-anthropocentric, and thus a generalization of this entailment relation at the level of Bedeutung rather than Sinn. We must therefore find an account of objective entailment, which will henceforth be called just 'entailment.'

As far as I am aware, there have been only two attempts to define entailment. The first is inspired by Tarski's definition (1935) of formal entailment, as a relation between sentences of a formal language.[8] One sentence Φ formally entails another, Ψ, if every interpretation that satisfies Φ also satisfies Ψ, roughly speaking. This idea is adapted to the problem of defining real entailment as follows: one state of affairs *A* entails (or includes) *B* just in case every possible world that satisfies *A* also satisfies *B*. The second attempt to define entailment is a more radical idea, due to Peter Gärdenfors (1988: 135). He introduces the notion of an epistemic state into logic, and defines propositions as functions between epistemic states. The proposition *A* then entails *B* just in case the composite function *AoB* equals the function *A*. This means, roughly, that *A* entails *B* just in case a rational thinker would infer *B* with certainty from *A*. We shall see that Gärdenfors's idea is by far the more fertile of the two.

Let us look first at the Tarski-style definition. The two key terms here are 'possible world,' and 'satisfaction,' so we must first be clear on what they mean. A possible world is a possible state of affairs, a way things might have been, but not every state of affairs is a possible world. A possible world has the special feature of being maximal, in some sense. The obvious difficulty here is that to be maximal is, apparently, not to be entailed by (or included in) any other state of affairs. The definition of a possible world thus requires a prior understanding of entailment between states of affairs. This is bad enough, but matters get even worse when we consider the satisfaction relation. What is it for a possible world to satisfy a state of affairs? A world *w* satisfies a state of affairs *A* just in case *A* necessarily obtains if *w* does. In other words, *w* satisfies *A* just in case *w* entails, or includes, *A*! We see that this approach to defining entailment is hopelessly circular.

Gärdenfors's foundation for logic is based upon the concept of an epistemic state. In Fregean spirit, he emphasizes that this is not a psychological notion, but is rather epistemological. It is a state of rational belief for an idealized epistemic agent. (An important part of the idealization is that every epistemic state is self-consistent and deductively closed.)[9] In his work, of analysing epistemic expansions and contractions, Gärdenfors looks at idealized human epistemic states, however, which are not suitable for our purpose. As stressed above, we are trying to understand entailment at the level of Bedeutung, as a relation between objective states of affairs. An examination of human epistemic states, however idealized, can only shed light upon entailment at the human level of Sinn, as a relation between Gedanken.

We will therefore consider the epistemic states of a perfect, infinite intellect, a mind of unlimited capacity that infallibly draws all and only valid inferences. This mind is an embodiment of all logical truth. I assume that the Gedanken of this being are (or are indistinguishable from) states of affairs themselves, so that the Sinn/Bedeutung problem does not arise. For this reason I will treat states of affairs as components of epistemic states, and definable from epistemic states, rather than as separate entities.[10] The analysis of states of affairs and the entailment relation roughly sketched below is my own, although it borrows heavily from Gärdenfors (1988: 132–45).

Ellis (1976) also offers an epistemic foundation for logic, although I make no use of it. Instead of ideal epistemic states, Ellis appeals to 'rational belief systems.' A belief system, which requires a language, specifies the attitude of a speaker (belief, disbelief, agnosticism) toward each sentence of the language. A belief system is rational just in case it is 'ultimately defensible before an audience of competent speakers' (1976: 201). An argument is valid, for Ellis, if there is no rational belief system on that language in which its premisses are accepted and its conclusion rejected. Thus, the relation of entailment is defined in terms of competency in speaking a language. Roughly speaking, a valid inference is one that a competent speaker would draw.

Ellis's system is similar to mine in this respect, as I define a valid inference (roughly) as one that would be drawn by a perfect mind. In other respects, however, our systems are quite dissimilar. The propositions in my system are objective states of affairs, rather than sentences, so that the whole subject matter is different. Also, the order in which he defines the terms is quite different from mine. Ellis defines a belief system using the sentences of a language, whereas for me the epistemic state is primitive. Propositions are defined on the structure of possible epistemic states. This difference forces many others, such as the way in which propositional operations are defined. My system is much closer to that of Gärdenfors, so let us begin by examining his ideas.

Gärdenfors is interested in changes of epistemic state with time, which fall into two basic kinds, called *expansions* and *contractions*. (A third kind of change, called a *revision*, may be understood as a contraction followed by an expansion.) An expansion occurs when one acquires extra knowledge, or learns new information. A contraction occurs when one discovers a mistake of some kind, and so gives up previously held beliefs. It is important to understand that when new information acts as a defeater for some previously held beliefs, reducing their probabilities, this is an expansion rather than a contraction. A contraction occurs only when one decides that the old beliefs were mistaken even on the information one had at the time. Both expansions and contractions are possible for human beings, but contractions can never occur for a perfect intellect, so we only need consider expansions.

The epistemic states can be seen as the points of a kind of logical space. States of affairs are present in this space, but not as independent entities. Rather, they are vectors that take you from one epistemic state to another. The rough idea is that a state of affairs is a carrier of information, and so the vector for a state of affairs $A$ represents the change of knowledge due to learning that $A$ is actual. In this picture, the fundamental logical relations are those between epistemic states, with relations between states of affairs defined in terms of them.

The chief logical relation between epistemic states, which is the root of the entailment relation, I call *superiority*. The rough idea of superiority is that a state $K$ is superior to $K'$, which we write $K \geq K'$, just in case $K$ knows at least as much as $K'$, that is, nothing is known in $K'$ that is not also known in $K$. In Gärdenfors's terminology, $K \geq K'$ just in case $K$ is an expansion of $K'$. Alternatively we may say that $K \geq K'$ if it is possible to move from $K'$ to $K$. Since knowledge cannot be lost, only gained, it is only possible to move from one state to a superior one. Note that each state is superior to itself, as a 'move' from $K$ to $K$ is trivially possible. If $K$ is superior to $K'$, then we shall say that $K'$ is inferior to $K$.

To define states of affairs in terms of epistemic states, we first need the notion of a maximal epistemic state. A maximal epistemic state is a state $K_w$ for which there does not exist any $K'$ (other than $K_w$ itself) such that $K' \geq K_w$. What, intuitively, does a maximal state look like? It should be a state of certainty about every possible state of affairs $A$; that is, one believes either $A$ or its negation with certainty. Such an epistemic state is a 'learning terminus': after one gets there, one can move nowhere else. It is a state that cannot be expanded further. $K_w$ is therefore the epistemic state where it is believed with certainty that some possible world, say $w$, obtains. It is shown below that possible worlds can be defined from maximal epistemic states.

A general state of affairs $A$ will be abstracted from the epistemic state $K_A$, where $K_A$ is the minimal state in which $A$ is fully believed. How can we define $K_A$ without invoking $A$? We first define a disjunction of epistemic states, as follows.

**2.2.2.1 Definition** The disjunction of states $\{K_1, K_2, ..., K_n\}$ is the maximal state that is inferior to each of $K_1, K_2, ..., K_n$. The disjunction of $K_1$ with $K_2$ is written $K_1 \vee K_2$.

This matches the usual truth-functional definition of disjunction, as the disjunction of two sentences is their strongest common consequence. (Note how there is a disjunction relation between epistemic states, which is ultimately used to define disjunction between states of affairs, and also that disjunction is defined in terms of superiority.) We then define pure epistemic states:

**2.2.2.2 Definition** (i) Every maximal epistemic state is pure.
(ii) The disjunction of a set of pure epistemic states is also pure.
(iii) No state is pure unless (i) and (ii) together entail that it is.

What, intuitively, is a pure epistemic state? It is generally recognized that a state of affairs is uniquely determined by the set of all possible worlds that include it. This is basically the idea that the meaning of a sentence is determined by its truth conditions, that is, the conditions under which the sentence is true. Now consider a pure state $K$ that is the disjunction of maximal states $K_1, K_2, ..., K_n$, corresponding to belief in worlds $w_1, w_2, ..., w_n$, and let the state of affairs $A$ be the disjunction of those worlds. Since each of $w_1, w_2, ..., w_n$ entails $A$, it follows that $A$ is believed with certainty in each of the states $K_1, K_2, ..., K_n$, and hence $A$ is believed with certainty in $K$. Now, is there any other state $K' \leq K$, such that $A$ is also believed with certainty in $K'$? Such a $K'$ would also be inferior to each of $K_1, K_2, ..., K_n$, since inferiority is transitive. Within the state $K'$ one is certain that one of the worlds $w_1, w_2, ..., w_n$ obtains, as these are the only worlds consistent with $A$. It follows from this, however, that $K' \geq K$, since in $K$ one merely knows (for certain) that one of the worlds $w_1, w_2, ..., w_n$ obtains. Thus $K' \geq K$, and $K' \leq K$, so $K' = K$, and we see that $K$ is the minimal state in which $A$ is believed with certainty. We can write this state as $K_A$. Each pure state can be used to define a state of affairs, as is shown below.

Before we can define states of affairs, we need some preliminary definitions.

**2.2.2.3 Definition** A class of epistemic states is consistent iff there is some epistemic state that is superior to all its members.

**2.2.2.4 Definition** The conjunction of $\{K_1, K_2, ..., K_n\}$ is the minimal state that is superior to each of $K_1, K_2, ..., K_n$. The conjunction of $K_1$ with $K_2$ is written $K_1$ & $K_2$.

Clearly, if the class $\{K_1, K_2, ..., K_n\}$ is not consistent, then these states do not have a conjunction.[11] This is in contrast to conjunction as an operation on sentences, where the conjunction always exists. The essential difference is that, while there are sentences ($\Phi$ & $\neg\Phi$, for example) that are inconsistent, there are no inconsistent epistemic states. Each epistemic state could be the epistemic state of an ideal intellect.

In Gärdenfors's account a state of affairs is a function, although he does not specify which one exactly. Since the function represents the expansion upon learning that $A$ obtains, it makes sense for the domain of $A$ to be the class of epistemic states consistent with $K_A$, and for $A$ to map each such state $K$ to $K \& K_A$.

**Definition** The Gärdenfors function $A$, whose domain is the set of epistemic states consistent with $K_A$, maps $K$ to $K \& K_A$. We write the image of $K$ under $A$ as $A(K)$.

Under this definition, $A$ entails $B$, that is, $K_A \geq K_B$, just in case $AoB = A$, so his proposed account of entailment works. We shall have occasional use for these Gärdenfors functions, but a better way to define states of affairs is as follows.

Consider some epistemic state $K$ that is consistent with $K_A$. In general, when the Gärdenfors function $A$ is applied to $K$, resulting in $K \& K_A$, the actual expansion (or learning) involved is rather less than $A$, since $K$ might already have some of the information carried by $A$. In the extreme case where $K \geq K_A$, for example, no learning takes place at all, as $A(K) = K$. If we want to talk about the actual learning that takes place on a particular expansion, therefore, the Gärdenfors function does not help.

One expansion where the full learning of $A$ takes place is if one moves to $K_A$ from an initial state of no information at all. We define this state as follows:

**2.2.2.5 Definition** The epistemic state $K_O$ is the disjunction of every epistemic state.

It is clear that $K_O$ is the minimal state, being inferior to every epistemic state. I assume that it is also equal to the disjunction of all the maximal states, and thus a pure state. In that case, it defines a state of affairs $O$, whose truth set is the class of all possible worlds. $O$, in other words, is the unique necessary state of affairs. Any sentence that is necessarily true, such as 'Hesperus is Phosphorus,' has $O$ as its Bedeutung. Again, we see a contrast between symbolic logic and real logic. In symbolic logic there are many tautologies, whereas in real logic there is only one 'tautology' $O$.[12] We now define states of affairs.

**2.2.2.6 Definition** The state of affairs $A$ is the expansion from $K_O$ to $K_A$.

**2.2.2.7 Corollary** The possible world $w$ is the expansion from $K_O$ to $K_w$.

We can now define entailment, and the usual logical connectives, between states of affairs.

**2.2.2.8 Definition** $A$ entails $B$, i.e., $A \Rightarrow B$, just in case $K_A \geq K_B$.

Since states like $K_A$ & $K_B$ are easily shown to be pure, we can define the usual logical connectives in the obvious way, namely:

**2.2.2.9 Definition** $K_{A\&B} = K_A \,\&\, K_B$.
$K_{A \vee B} = K_A \vee K_B$, etc.

**2.2.2.10 Definition** Let $t(K)$, the truth set for $K$, denote the class of possible worlds consistent with $K$. (Note that $K$ may be any epistemic state, and does not have to be pure.)

**2.2.2.11 Theorem** If $K' \geq K$, then $t(K') \subseteq t(K)$.

*Proof*: Suppose that $K' \geq K$, but that for some world $w \in K'$, $w \notin K$. Then $K$ knows that $w$ is not the case, whereas $K'$ does not know this. Thus, contrary to the hypothesis, $K'$ is not superior to $K$.■

**2.2.2.12 Corollary** If $A$ is believed with certainty in $K$, and $K' \geq K$, then $A$ is believed with certainty in $K'$.

*Proof*: Since $A$ is certain in $K$, it follows that $A$ is entailed by each member of $t(K)$. Then, since $K' \geq K$, it follows that $t(K') \subseteq t(K)$, and so $A$ is entailed each member of $t(K')$. Since $A$ obtains in every world consistent with $K'$, $A$ must be certain in $K'$.■

**2.2.2.13 Theorem** $A$ entails $B$ iff $B$ is believed with certainty in $K_A$.

*Proof*: (i) If $A$ entails $B$, then $K_A \geq K_B$. $B$ is believed with certainty in $K_B$, and hence in $K_A$ as well, since it is superior to $K_B$. (ii) If $B$ is believed with certainty in $K_A$, then $K_A \geq K_B$.■

We see that the definition of $A \Rightarrow B$, that $K_A \geq K_B$, is equivalent to the statement that $B$ is certain in $K_A$. Now, at the human level, belief is a matter of degree, so that there are grades of certainty with regard to a state of affairs, ranging from firm belief down to firm disbelief. A human epistemic state is not dogmatic with regard to every state of affairs; on some matters one is sceptical, tentative, or unsure. It is plausible to suppose, therefore, that the same is true of ideal epistemic states. Consider, for instance, the state of affairs $A$, that Venus has either one moon or none, and $B$, that Venus has no moon. Within $K_A$, is $B$ believed or not? It would be irrational for $B$ to be believed with certainty, and also for $\neg B$ to be believed with certainty, as neither of these is entailed by $A$. The correct attitude within $K_A$ is surely some partial belief in $B$. Unless epistemic states can include such partial beliefs, it is impossible to regard them as a rational ideal.

If an epistemic state can include partial beliefs, however, then there is a natural way to generalize the entailment relation to include degrees of entailment.[13]

**2.2.2.14 Definition** $A$ entails $B$ to degree $p$ iff $B$ is believed to degree $p$ in $K_A$, i.e., $Pr(B \mid A)$ is the degree to which $B$ is believed in $K_A$.

### *2.2.3 Truth and Authority*

One of the maximal epistemic states, $K_T$ say, has a rather special relationship with the real, concrete world. It does not make sense to say that $K_T$ is true, since it is itself the standard for truth and falsehood. The actual world, $T$, is merely the expansion from $K_O$ to $K_T$, and the real world is an embodiment, or concrete version, of $T$. $K_T$ is the formal cause of the real world, and the defining characteristic of true sentences and beliefs, so we may call it the truth. It might seem a little extreme to define the truth as an epistemic state,[14] but it is the only way I can see to understand the relation between belief and truth. The truth/belief distinction is just the Bedeutung/Sinn distinction. (It is interesting that postmodern thinkers, who reject the very notion of truth, use 'truth' and 'belief' as synonyms, talking about 'your truth' and so on.)

**2.2.3.1 Definitions**
(i) $K$ is *veridical* iff it is consistent with $K_T$.
(ii) A state of affairs $A$ is *actual* just in case $K_A$ is veridical.
(iii) A *true* sentence is one whose Bedeutung is an actual state of affairs.

Suppose one knows for sure that the state of affairs $A$ obtains. Moreover, $A$ in fact entails some other state of affairs $B$. One may well not believe that $B$ obtains, as one fails to see that $B$ follows from $A$. Perhaps one still cannot see the entailment even when it is pointed out and explained. If, however, one somehow learns that $A$ does include $B$, then one can infer $B$ from these two premisses: $A$ and $A \Rightarrow B$. (This is quite a different inference from deducing $B$ directly from $A$ as a single premiss.) In the two-premiss inference, one is effectively submitting to the authority of logic, accepting its verdict, whereas in the one-premiss inference one does the thinking for oneself. Both inferences are valid, of course.

Logical relations of all kinds seem to be authoritative in this way with regard to belief. If one learns that two states of affairs are inconsistent, for example, then one should not believe firmly in both of them. It is reasonable to suppose, therefore, that logical probabilities are also authoritative.

**2.2.3.2 Definition** If $K \geq K'$, and $K$ is veridical, then $K$ is an authority for $K'$.

This is a rather stringent definition of relative authority; I imagine that some weaker condition would be sufficient, but this definition will serve our purpose.

**2.2.3.3 Principle** (Authority of Logic)

If one knows that $K$ is an authority for one's epistemic state, and knows that $A$ is believed to degree $p$ in $K$, then one is (defeasibly) authorized to believe $A$ to degree $p$.

This is the foundation for Miller's Principle, which I call the Authority of Chance principle. The authorization is clearly defeasible, as there may be more than one such authority. The only non-defeasible authorization occurs when $p = 1$, so that $A$ is known with certainty.

### *2.2.4 Relative States of Affairs*

We defined a state of affairs $A$ as the expansion from $K_O$ to $K_A$. Of course, $K_O$ is not the only epistemic state that may be expanded upon learning that a state of affairs obtains. For instance, there is the smaller expansion from $K_U$ to $K_{A\&U}$.

**2.2.4.1 Definition** The relative state of affairs $A/U$, read '$A$ given $U$,' is the expansion from $K_U$ to $K_{A\&U}$.

Two distinct states of affairs, $A$ and $B$, may be such that, for some third state of affairs $U$, $A/U = B/U$, that is, $A(K_U) = B(K_U)$, where $A$ and $B$ are the Gärdenfors functions. For example, if $U$ is Smith's being an Albertan, $A$ is Smith's being a farmer, and $B$ is Smith's being a Canadian farmer, then $A/U$ and $B/U$ are the same relative state of affairs.

The conjunction operation, for absolute states of affairs, is defined above in Def. 2.2.2.9. This definition cannot be used for relative states of affairs, since there is no epistemic state $K_{A/U}$, in general. The expansion $A/U$ cannot be applied to the state $K_O$, I think, because $A/U$ presupposes $U$ – it simply makes no sense from the point of view of $K_O$. For example, the relative state of affairs that the Loch Ness Monster weighs 300 tons only makes sense relative to the state of affairs that there is a monster in Loch Ness.

In the case of absolute states of affairs, Def. 2.2.2.9 is equivalent to the following:

**2.2.4.2 Definition** $K_{A\&B} = A(K_B) = B(K_A)$, where $A$ and $B$ are Gärdenfors functions.

Now, this idea can be applied to relative states of affairs, in cases where the expansion $A/U$ can be applied to a state $K_C$. Such cases occur when the presuppositions made by $A/U$ are all contained in $K_C$. For instance, if $A$ is logically independent of $U$, then $A/U$ presupposes nothing, and can be applied to any epistemic state. Also, $A/U$ can be applied to $K_U$, and any state superior to $K_U$.

**2.2.4.3 Definition** $K_{C\&(A/U)}$ is the state obtained from applying $A/U$ to $K_C$.

**2.2.4.4 Theorems** (i) $B \,\&\, A/B = A \,\&\, B$.
(ii) If $A \Rightarrow B$, then $A = B \,\&\, A/B$.

*Proof*: Since $A/B$ is defined as the expansion from $K_B$ to $K_{A\&B}$, part (i) is trivial. Part (ii) follows from (i), since if $A \Rightarrow B$ then $A \,\&\, B = A$.■

Using the idea of a relative proposition, we can define the logical independence of two states of affairs as follows.

**2.2.4.5 Definition** $A$ and $B$ are *logically independent* in $K_U$ just in case $A/(B \,\&\, U) = A/U$.

It is useful to define logical probabilities for relative propositions, that is, give a meaning to the symbol '$Pr(A/B)$.' Within the state $K_B$, the relative proposition $A/B$ contains the same information as the absolute proposition $A$. Thus, it makes sense for $Pr(A/B \mid B)$ to equal $Pr(A \mid B)$.

**2.2.4.6 Theorem** If $A$ and $B$ are logically independent within $K_U$, and $Pr(A \mid B \& U)$, $Pr(A \mid U)$ exist, then $Pr(A \mid B \& U) = Pr(A \mid U)$.

Relative states of affairs are required later in this chapter, in the definition of symmetry between states of affairs, and again in chapter 6, where they are essential to the interpretation of the quantum-mechanical state vector.

## 2.3 MEASURING DEGREES OF BELIEF

Having sketched out the nature of logical probability, let us now proceed to giving a precise account of it. The rough definition we have so far is: $Pr(A \mid B)$ is the degree to which $A$ is believed in $K_B$. So the first task is to devise a system for measuring degrees of belief.

One standard way to define degrees of belief is using gambles, but I prefer to use contracts instead, as they are more flexible. A contract is something like this: I'll give you $g$ if it rains tomorrow. The general form is [$g$ if $A$], where $g$ is some desired thing and $A$ is a state of affairs. If you possess that contract, then you acquire $g$ if $A$ is true, and nothing (from that contract) otherwise. It is easy to define a gamble on $A$, with betting quotient $p$ and stake \$1, as the purchase of [\$1 if $A$] at the price \$$p$.

If the object $g$ is desirable, then clearly the contract [$g$ if $A$] has some value too, as whoever possesses it might get $g$ from it. Moreover, the worth of the contract depends on the degree to which one believes $A$ – indeed, the value of the contract increases strictly with the degree of belief in $A$. Thus, we can use the value of the contract to measure the degree of belief, that is, to give 'degree of belief' a precise meaning. We shall therefore be defining the probability of $A$ given $B$ as the value of a contract.

It may seem odd to consider desires and preferences as part of logic, but these need not be human preferences. Indeed, it surely makes sense to use the desires and preferences of the perfect, infinite mind already invoked in §2.2. If this being has beliefs, then why not desires as well? Moreover, we shall see that to define values for arbitrarily great goods, which are needed to define real-number values, one must consider the preferences of a being of infinite capacity.

It should be noted that this mind is not assumed to be omniscient, that is, its epistemic state need not be maximal. If it were necessarily omniscient, then all its possible epistemic states would be possible worlds, and so all logical probabilities would be either 0 or 1. This mind is assumed to be 'logically omniscient,' on the other hand, in the sense that it knows the logical consequences of the propositions it knows.

We require a function, which I call *val*, that maps each contract to a real number that measures its value, within an epistemic state. What meaning can this number have? In general, one measures[15] objects by reference to a standard object, called a unit. Mass, for example, is measured by reference to a standard massive object, as the number of instances of the unit required to equal the body in question with respect to mass. In a similar way, the rough idea of $val(g)$ is the number of units required to equal $g$ with respect to preference. Let us begin by describing the different preference relations.

The fundamental preference relation between desirable objects $a$ and $b$ is that a is preferred to $b$, which is written $a \geq b$.

**2.3.1 Definition** $a \geq b$ just in case there is no objection to giving up $b$ in return for $a$.

Note that the relation is reflexive, that is, $a$ is preferred to $a$. This relation of preference can be used to define three further preference relations, as follows.

**2.3.2 Definition**
(i) $a > b$, i.e., $a$ is *strictly* preferred to $b$, iff $a \geq b$ but not $b \geq a$.
(ii) $a \sim b$, i.e., $a$ and $b$ are preference *equivalent*, iff $a \geq b$ and $b \geq a$.
(iii) $a <> b$, i.e., $a$ and $b$ are preference *incommensurable*, iff neither is preferred to the other.

Let us denote the unit desired thing by '\$'. (We will call it 'dollar,' but its true identity is not relevant.) In order for it to be suitable as a unit, it must be reproducible, so that there can be arbitrarily many other objects $\$_1$, $\$_2$, ..., all similar to \$, so that $\$_i \sim \$_j$. One may possess several of these dollars, in which case we shall say that one owns the bundle $\{\$_1, \$_2, ...\}$. A bundle is a simple collection, or aggregate, rather than a set, so that a singleton bundle is equal to its only member, $\{\$_1, \{\$_2, \$_3\}\} = \{\$_1, \$_2, \$_3\}$, and so on.

If the dollar is to work as a unit of value, it is clearly necessary that equal-sized bundles of dollars be preference equivalent – e.g., $\{\$_1, \$_2\} \sim \{\$_3, \$_4\}$ as otherwise a cardinal number of dollars would not specify a unique value. In that case one could not say that an object is worth two dollars, without saying which two dollars! We thus postulate that, for any bundle of dollars *a*, $\{\$_i, a\} \sim \{\$_j, a\}$. Assuming the dollars are all intrinsically similar, this is sure to hold. We will henceforth denote a bundle of *n* dollars by '$\$n$', as the identity of each dollar is irrelevant.

Now let us consider contracts. It makes sense to use contracts for dollars, such as [$\$r$ if *A*], as the object $\$r$ always has a precise value, namely, *r*. Indeed, from now on the term 'contract' will always mean a contract of the form [$\$r$ if *A*]. In order for the value function to be linear, in the sense that the value of a bundle (of dollars and contracts) equals the sum of the values of its parts, we need to assume a principle known as independence, as follows.

**2.3.3 Principle** (Independence)

If [$\$r$ if *A*] ~ $\$q$, then {[$\$r$ if *A*], *a*} ~ {$\$q$, *a*}, where *a* is any bundle of dollars and contracts.

This principle will not hold in general, as the following two examples show. First consider a coffee shop that gives away one $ with each cup of coffee bought there. When ten of these have been collected, they may be exchanged for a cup of coffee. Suppose further that each $ has little or no value in itself, but is only useful as a means for acquiring coffee. In this case, the principle of independence will fail. For suppose that the contract [$6 if *A*] is worth $3, for someone who has no other dollars. Then, if one already has $4, the contract [$6 if *A*] will be worth more than $3, as $10 is worth a cup of coffee whereas $7 is practically worthless.

Second, suppose that [$1000 if *A*] is worth $1. It does not follow that 1000 such contracts together (equivalent to the single contract [$1,000,000 if *A*]) is worth $1000, since it may be that $1,000,000 is more dollars than a person could possibly use in their whole lifetime. In that case one would be foolish to give up $1000 in return for the contract, as $1,000,000 may be worth only about twice as much as (say) $5000.

To avoid the first kind of example, we shall suppose that dollars are desired for their own sake, and for no other reason. Their value is purely intrinsic, so that (intuitively speaking) the value of a bundle of dollars is just the sum of the values of the parts. The second kind of example is actually impossible since we are considering not human desires, which may be saturated, but those of a being of infinite capacity. For such a being, the possession of *n* dollars (for any *n*) does not even slightly reduce the desirability of a further dollar.

If [$\$r$ if *A*] ~ $\$n$, then we define *val*[$\$r$ if *A*] = *n*. What about contracts that are not preference equivalent to a bundle of dollars, however? Fortunately, this value function is easily enriched to include quotients as well. If two *bs* together are worth \$1, for example, then each $b$ has value $\frac{1}{2}$, and three $b$s have joint value $\frac{3}{2}$, and so on. Finally, we can extend the value scale to the positive reals, using the method of Dedekind cuts. For a contract $a$ may be such that, for the value scale defined by \$, for every possible quotient value $r$, either $r < a$ or $r > a$.[16] The object $a$ then defines a Dedekind section on the \$-value quotient scale, and this cut may itself be regarded as a further possible value on the scale, represented by the obvious irrational number.[17] The precise definition of *val* is as follows, where $a$ is bundle of dollars and contracts. Note that $\{\}$ is an empty bundle – literally nothing at all.

**2.3.4 Definition** The value of $a$, or *val*($a$), is defined for the unit \$ as follows.

I If $a \sim \{\}$, then $val(a) = 0$

II If $\{a_1, ..., a_q\} \sim \$p$, then $val(a) = p/q$.

III For each object $a$, let $Q_* = \{r \in \mathbb{Q}:(\exists x)(val(x) = r \;\&\; x \leq a\}$ and $Q^* = \{r \in \mathbb{Q}:(\exists x)(val(x) = r \;\&\; x > a\}$. Assuming each quotient is the value of some object, $\langle Q_*, Q^*\rangle$ is a Dedekind cut. We then define *val*($a$) as the real number represented by $\langle Q_*, Q^*\rangle$.

Once again, it is quite clear that we are not concerned with human beings here. It is not plausible that desirability for humans is as finely graduated as this. Indeed, why suppose that it even makes sense to talk of such precise values? There is no a priori guarantee that it does make sense,[18] but let us see what follows from it. It should be noted that, according to Def. 2.3.4, real-number values are not precise, in the sense that *val*($a$) may equal *val*($b$) even if $a$ and $b$ are not preference equivalent. For instance, any object whose value is 'infinitesimal' will have the real number zero as its value, even though it is strictly preferred to the empty bundle $\{\}$.[19]

Although the equality of *val*($a$) and *val*($b$) does not entail that $a \sim b$, unless *val*($a$) and *val*($b$) are quotients, the converse does hold.

**2.3.5 Theorem** If *val*($a$) exists, and $a \sim b$, then $val(a) = val(b)$.

*Proof*: Suppose *val*($a$) exists and that $a \sim b$; then *val*($a$) is either a rational or an irrational number. Case (i): suppose that *val*($a$) is rational. Then, for some natural numbers $p$ and $q$, we have that $\{a_1, ..., a_q\} \sim \$p$. Now, applying independence $q$ times, together with the assumption that $a \sim b$, it follows that $\{b_1, ..., b_q\} \sim \$p$, and so $val(b) = p/q = val(a)$. Case (ii): *val*($a$) is irrational. It is clear that $a$ and $b$ will have the same preference relations to all objects of rational value; that is, if $a > x$ then $b > x$ and so on, and so $a$ and $b$ define the same Dedekind cut of the rationals. Thus $val(a) = val(b)$.■

## 2.4 THE AXIOMS OF PROBABILITY

We are now almost ready to demonstrate the four axioms of probability,[20] but first we shall need some definitions and preliminary results. The objects $a$, $b$, and so on are bundles of dollars and contracts. It should be noted that the proofs given below are not 'Dutch book' arguments, involving the dubious assumption that a set of betting quotients is irrational if it leaves one vulnerable to a Dutch book,[21] but are direct demonstrations.

**2.4.1 Definition** $Pr(A \mid B) = val[\$1 \text{ if } A]$ within $K_B$.

**2.4.2 Lemma** $val\{a, b\} = val(a) + val(b)$, in any state $K$.

*Proof*:
Case (i): $val(a)$ and $val(b)$ are natural numbers, say $p$ and $q$ respectively. We then have, by the definition of $val$, that $a \sim \{\$_1, \$_2, ..., \$_p\}$ and $b \sim \{\$_1, \$_2, ..., \$_q\}$. By independence it follows that $\{a, b\} \sim \{\$_1, \$_2, ..., \$_{p+q}\}$, and thus $val\{a, b\} = p + q$, as required.

Case (ii): $val(a)$ and $val(b)$ are rational numbers, say $p/q$ and $r/s$. Then, again by the definition of $val$, we have $\{a_1, a_2, ..., a_q\} \sim \$p$ and $\{b_1, b_2, ..., b_s\} \sim \$r$. Then, using Lemma 2.4.2 in Case (i), we have $\{a_1, a_2, ..., a_{sq}\} \sim \$sp$ and $\{b_1, b_2, ..., b_{sq}\} \sim \$qr$. Again using Lemma 2.4.2 in Case (i) it follows that $\{\{a_1, b_1\}, ..., \{a_{sq}, b_{sq}\}\} \sim \$(sp + qr)$. Thus, by definition, $val\{a, b\} = (sp + qr)/sq = p/q + r/s$, as required.

Case (iii): $val(a)$ and $val(b)$ are real numbers $r$ and $s$, represented by Dedekind cuts $\langle R_*, R^*\rangle$ and $\langle S_*, S^*\rangle$ respectively. Let $t = r + s$, and $t$ be represented by the cut $\langle T_*, T^*\rangle$. Then we are required to prove that $val\{a, b\}$ is represented by $\langle T_*, T^*\rangle$, that is, that $T_*$ is the set of quotient values of objects inferior to $\{a, b\}$, and $T^*$ is the set of quotient values of objects preferable to $\{a, b\}$. Consider arbitrary $t^- \in T_*$. It is always possible (by the definition of real addition) to find quotients $r^-$, $s^-$ in $R_*$, $S_*$ such that $r^- + s^- = t^-$; then select objects $a^-$, $b^-$ whose (quotient) values are $r^-$ and $s^-$. Then, by Lemma 2.4.2 Case (ii), $val\{a^-, b^-\} = r^- + s^-$, and it is also clear that $\{a^-, b^-\} < \{a, b\}$. Thus, each element of $T_*$ corresponds to an object inferior to $\{a, b\}$, and it may similarly be shown that each member of $T^*$ corresponds to an object preferable to $\{a, b\}$.■

**2.4.3 Lemma** $Pr(A \mid B) = p$ iff ([\$1 if $A$] ~ \$$p$) within $K_B$.

*Proof*: Immediate from Definition 2.4.1.■

**2.4.4 Lemma** [\$$p$ if $O$] ~ \$$p$, in any state $K$.

*Proof*: $K$ is superior to $K_O$, so $O$ is believed with certainty in $K$.■

**2.4.5 Lemma** *val*[\$$rp$ if $A$] = $r$.*val*[\$$p$ if $A$], in any state $K$.

*Proof*:
Case (i): $r$ is a natural number. Since [\$$rp$ if $A$] ~ {[\$$p$ if $A]_1$, ..., [\$$p$ if $A]_r$}, we can apply Lemma 2.4.2 to get *val*[\$$rp$ if $A$] = *val*[\$$p$ if $A]_1$ + ... + *val*[\$$p$ if $A]_r$.■

Cases (ii), where r is a quotient, and (iii) where r is real, may be proved using arguments similar to those given in the proof of Lemma 2.4.2.

**2.4.6 Lemma** If *val*[\$$p$ if $A$] = *val*[\$$q$ if $B$], then *val*[\$$rp$ if $A$] = *val*[\$$rq$ if $B$].

*Proof*: Immediate from Lemma 2.4.5.■

The axioms of the probability calculus may be shown as follows. The 'within $K_B$' part is omitted unless it is manipulated in the proof.

**Axiom 1** $Pr(A \mid B) \geq 0$

*Proof*: The contract [\$1 if $A$] either yields \$1 or nothing. It has to be worth at least zero therefore.■

**Axiom 2** $Pr(O \mid B) = 1$.

*Proof*: [\$1 if $O$] ~ \$1, from Lemma 2.4.4, and *val*(\$1) = 1.■

**Axiom 3** $Pr(A \vee B \mid C) = Pr(A \mid C) + Pr(B \mid C)$, where $A$ and $B$ are inconsistent within $K_C$.

*Proof*: *val*[\$1 if $A \vee B$] = *val*{[\$1 if $A$], [\$1 if $B$]}
= *val*[\$1 if $A$] + *val*[\$1 if $B$], using Lemma 2.4.2.■

**Axiom 4** If *Pr*(*B* | *C*) exists and is non-zero, then *Pr*(*A* & *B* | *C*) = *Pr*(*A* | *B* & *C*)*Pr*(*B* | *C*).

*Proof*:

| | |
|---|---|
| 1 ([\$1 if *A*] ~ [\$*Pr*(*A* \| *B*) if *O*]) within $K_B$ | Consequence of Lemma 2.4.3. |
| 2 ([\$1 if *A*] ~ [\$*Pr*(*A* \| *B* & *C*) if *O*]) within $K_{B\&C}$ | Consequence of Lemma 2.4.3. |
| 3 ([\$1 if *A* & *B*] ~ [\$*Pr*(*A* \| *B* & *C*) if *B*]) within $K_{B\&C}$ | From 2. Since *B* is known to be true, it can be added to both sides. |
| 4 ([\$1 if *A* & *B*] ~ [\$*Pr*(*A* \| *B* & *C*) if *B*]) within $K_{\neg B\&C}$ | Both contracts are worthless. |
| 5 ([\$1 if *A* & *B*] ~ [\$*Pr*(*A* \| *B* & *C*) if *B*]) within $K_C$ | From 3 and 4. One of $K_{B\&C}$, $K_{\neg B\&C}$ is an authority for $K_C$. |
| 6 ([\$1 if *B*] ~ [\$*Pr*(*B* \| *C*) if *O*]) within $K_C$ | Consequence of Lemma 2.4.3. |
| 7 ([\$*Pr*(*A* \| *B* & *C*) if *B*] ~ [\$*Pr*(*A* \| *B* & *C*) *Pr*(*B* \| *C*) if *O*]) given *C* | From 6, multiplying both sides by *Pr*(*A* \| *B* & *C*), using Lemma 2.4.6. |
| 8 ([\$1 if *A* & *B*] ~ [\$*Pr*(*A* \| *B* & *C*) *Pr*(*B* \| *C*) if *O*]) given *C* | From 5, 7, using transitivity. |
| 9 *Pr*(*A* & *B* \| *C*) = *Pr*(*A* \| *B* & *C*)*Pr*(*B* \| *C*) | From 8, by L. 2.4.3, as required.■ |

This fourth axiom of the probability calculus is also known as the Principle of Conditioning. Note that it is not a diachronic constraint on one's beliefs, for time has nothing to do with it; rather, it describes the partial entailment relation between states of affairs. It is often written in the equivalent form (suppressing *C* for simplicity): *Pr*(*A* | *B*) = *Pr*(*A* & *B*)/*Pr*(*B*), and referred to as the *definition* of the symbol *Pr*(*A* | *B*), or as the definition of 'conditional probability.' This is a mistake, for two reasons. First, all probabilities exist only relative to an epistemic state, so *Pr*(*B*) means *Pr*(*B* | *O*), which does not exist in general. Second, the symbol *Pr*(*A* | *B*) already has a meaning, as the value of

[$1 if $A$] within $K_B$. That is why we read it as 'the probability of $A$ *given* $B$.' One cannot give two separate definitions for the same symbol, unless their equivalence is demonstrated.

## 2.5 RELATIVE PROBABILITIES

It is common to talk of relative probabilities, such as in the statement '$A$ is three times as likely as $B$.' What do these assertions mean, however? Perhaps the most obvious answer is that '$A$ is three times as likely as $B$' means that $Pr(A)/Pr(B) = 3$, or

$$Pr(A) = 3.Pr(B) \qquad (1).$$

Using Definition 2.4.1, (1) is equivalent to $val$[$1 if $A$] = 3.$val$[$1 if $B$], or

$$val[\$1 \text{ if } A] = val[\$3 \text{ if } B] \qquad (2).$$

Statement (2) entails, but is not equivalent to,

$$[\$1 \text{ if } A] \sim [\$3 \text{ if } B] \qquad (3).$$

Although (1) entails (3), statement (3) does not entail (1) since $val$[$1 if $A$] and $val$[$3 if $B$] might not exist. Thus (3), being weaker than (1), requires a new notation, and will be written $R(A, B) = 3$. In general we have the following definition.

**2.5.1 Definition** For propositions $A$, $B$ we define $R(A, B)$, the probability of $A$ relative to $B$, as follows: $R(A, B) = p$ just in case [$1 if $A$] ~ [$$p$ if $B$].

The following theorems are obvious.

**2.5.2 Theorem** $Pr(A) = R(A, O)$.

**2.5.3 Theorem** If $Pr(B)$ exists, then $R(A, B) = Pr(A)/Pr(B)$.

A more interesting theorem is a generalized version of the principle of conditioning, which is proved in the same way as Axiom 4 (although more easily).

**2.5.4 Theorem** $R(A, B \mid U) = R(A \,\&\, U, B \,\&\, U)$

**2.5.5 Corollary** *Pr*(*A* | *U*) = *R*(*A* & *U*, *U*)

The advantage of these results over Axiom 4 is that there is no need for *Pr*(*U*) to exist. This is of particular importance to cases where one wants to condition on very unlikely events, whose probability might be described as 'infinitesimal.'

Consider, for example, an infinite lottery, where a ticket is selected at random from an urn that contains a countably infinite number of such tickets. Does it make sense to assert that every ticket has the same probability of being selected? Though this seems like a meaningful scenario, one sometimes hears that it is impossible, on the grounds that it cannot be represented by a probability function. If each ticket has probability δ, then δ has to be either zero or a finite, positive number. Both of these are ruled out by the addition axiom, however. The problem here is that, if each ticket has the same probability, then that probability has to be infinitesimal, so to speak. If the proposition *A* says that some ticket *a* is picked, then [\$1 if *A*] > \$0, since *A* might be true, but [\$1 if *A*] < \$δ, for each δ > 0.

Using relative probabilities, we can express the fact that *A* and *B* are equally likely by the statement *R*(*A*, *B*) = 1. Thus, if *b* is some ticket distinct from *a*, and *B* says that *b* is picked, then it is meaningful to assert that [\$1 if *A*] ~ [\$1 if *B*], that is, *R*(*A*, *B*) = 1.

It is clear that the 'absolute' probability function *Pr* can be reduced to the relative probability function *R*, since *Pr*(*A* | *B*) = *R*(*A*, *O*), within the state $K_B$. Is a reduction possible in the reverse direction, however? Can relative probabilities be reduced to relations between absolute probabilities? The answer to this is clearly No, if we are restricted to real-number probabilities, as the case of the infinite lottery shows. This limitation of the real-number system is surely related to the point made in connection with Definition 2.3.4, that real numbers are only approximate representations of measure. Two objects of slightly different value can generate the same Dedekind cut on the quotients. Unfortunately there does not seem to exist any system of measures that is more accurate than the real numbers.

## 2.6 INTERVAL PROBABILITIES

In §2.4 we assumed that the contract [\$1 if *A*] had an exact dollar value within $K_B$. This is not the case for every pair of propositions *A* and *B*, however, as we shall see below. More generally there exists some interval of values, each of which is incommensurable with [\$1 if *A*].

Suppose that, for some good \$*p*, we have that [\$1 if *A*] <> \$*p*. This entails that there does not exist any good $*a* such that [$1 if *A*] ~ $*a*, so that *val*[$1 if *A*] does not exist, as the following theorem shows.

**2.6.1 Theorem** If [$1 if *A*] <> $*p*, then there is no good $*a* such that [$1 if *A*] ~ $*a*.

*Proof*: Suppose that there is some *a* ∈ [0,1] such that [$1 if *A*] ~ $*a*. Then, since *p* is also a real number, we have that either *p* < *a*, *p* = *a*, or *p* > *a*. These cases entail, however, that [$1 if *A*] > $*p*, [$1 if *A*] ~ $*p*, and [$1 if *A*] < $*p* respectively, contrary to the hypothesis.■

**2.6.2 Definition** If *val*[$1 if *A*] does not exist, within $K_B$, then let *Pr*(*A* | *B*) equal the set *I*, where *I* = {*p*: [$1 if *A*] <> $*p* (within $K_B$)}.

**2.6.3 Theorem** *I* is an interval (if we include ∅ as an interval).

*Proof*: Any set *D* such that, if *a*, *c* ∈ *D* and *a* < *b* < *c*, then *b* ∈ *D* is an interval. (Thus, the null set and singletons are intervals.) For the sake of reductio, therefore, let a, *c* ∈ *I*, but suppose that there exists some *b* ∉ *I* in between them, that is, *a* < *b* < *c*. Since *b* ∉ *I*, it follows that either [$1 if *A*] < $*b*, [$1 if *A*] ~ $*b*, or [$1 if *A*] > $*b*. In each of these cases we can refute at least one of the statements *a* ∈ *I*, *c* ∈ *I*. Thus, by reductio, each *b* between *a* and *c* is a member of *I*, and so *I* is an interval.■

Thus, if we define *Pr*(*A* | *B*) as in Def. 2.6.2, a probability is a sub-interval of [0,1]. One kind of example where *Pr*(*A* | *B*) is plausibly an interval is the following. Suppose *B* says that, in a particular urn, the proportion of black balls is somewhere in the interval [*a*, *b*], and that a ball is drawn from the urn at random. *A* says that the ball drawn is black. Let us assume for now that, if the proportion of black balls were known to be *p*, then the probability would also be *p*. It then follows that *Pr*(*A* | *B*) = [*a*, *b*] in this example.

## 2.7 THE SYMMETRY AXIOM

As I have presented them, the axioms of probability are not part of the definition of the logical probability function, for that is defined in Def. 2.4.1. Moreover, within the theory of logical probability they do not function as axioms, but as theorems, since they are proved from the definition of probability. The purpose of proving the axioms is not to define probability, nor to argue that the axioms are true, but rather to confirm Def. 2.4.1 by showing that the axioms all follow from it. An analysis of logical probability that did not allow one to prove the axioms would be unacceptable.

One common element in previous accounts of logical probability is the use of a symmetry, or indifference, principle. It is widely felt that, if there are logical probabilities, then they must be assigned largely (though by no means wholly) on the basis of symmetries. I also hold this conviction, and so in this section I shall develop and prove a fifth axiom, the symmetry axiom. The notion of symmetry involved in the axiom is very strict, however, so that the axiom is of limited application. In most of the cases where one of the standard indifference principles can be applied, there is no symmetry of the sort required to use the symmetry axiom. The purpose of developing the symmetry axiom is philosophical rather than practical; I am proving the point that, in some cases at least, a purely logical principle can determine exact rational degrees of belief.

While the notion of symmetry needed here is a *logical* concept, rather than a geometrical one, the notion of symmetry is better known in geometry, so we shall begin there. The general idea of geometrical symmetry is that a figure is mapped to itself under some transformation that is not the identity transformation. In other words, if $A$ is a plane figure and $f$ some transformation (other than identity), then $A$ has $f$-symmetry just in case $fA = A$. For example, an equilateral triangle has symmetry with respect to a rotation of 120° about its centre. The function $f$ cannot be just any mapping, however, but must be an *isometry*, that is, a transformation that preserves the distance between each pair of points. Although some figures map to themselves under functions that are not isometries, this fact does not make those figures symmetric. In general, we shall use the term *symmetry function* to refer to a function $f$ that makes a figure $A$ symmetric just in case $fA = A$.

To make this idea precise, we shall represent a plane 'universe' $U$ as a 'colouring function' $u$, which maps $\mathbb{R}^2$ to $\{0, 1\}$. The idea is that $u(x, y)$ represents the colour of the point $(x, y)$, where 0 means the background colour (white, say) and 1 means the ink colour, perhaps black. In this representation a figure is a set of points that are coloured black. A transformation on $U$ is a function $f$ that maps $\mathbb{R}^2$ to $\mathbb{R}^2$, as usual. What colouring function $u'$ represents $fU$, that the image of the universe $U$ under $f$? If $u(x, y) = 1$, then we want that black point to be moved to $f(x, y)$, so that $u'(f(x, y)) = 1$. White points need to be moved in the same way. We therefore have that $u'(f(x, y)) = u(x, y)$, which entails that $u' = uf^{-1}$. Thus, if $u$ represents $U$, then $fU$ is represented by $uf^{-1}$.

Thus, a figure $A$ has $f$-symmetry, where $f$ is a symmetry function, just in case $fA = A$. We will now extend this idea to define a *relation* of symmetry, that is, to say what it means for *two* figures to be symmetric. What could it mean, for instance, to say that the pair $\{A, B\}$ has reflective symmetry?

According to the definition of a figure given above, as a set of points that is coloured black, a pair of figures such as $\{A, B\}$ defines a third, joint figure $A \cup B$, the set of (black) points that are in either $A$ or $B$. We then have the following definition.

**Definition** $A$ has $f$-symmetry with $B$ just in case $A \cup B$ has $f$-symmetry, and $fA = B$.

One kind of case where $A$ has $f$-symmetry with $B$ can be constructed as follows. Consider, for instance, a figure $A$ that is not $f$-symmetric, and suppose that the function $f$ is its own inverse, that is, $f = f^{-1}$, or $ffA = A$.[22] Then the joint figure, $A \cup fA$, has $f$ symmetry in the above sense. This is clear since $f(A \cup fA) = fA \cup ffA = fA \cup A$, as required. Thus, if $B$ is $fA$, then $A$ has $f$-symmetry with $B$.

Figure 2.1

Suppose that $A$ and $B$ are $f$ symmetric, and consider a plane that contains only the figures $A$ and $B$. This plane will be called the universe $U$. It is clear that, within $U$, there is no way to pick out either $A$ or $B$ using only distances. If someone asks which of the figures is $A$, and which is $B$, one cannot say: '$A$ is the one that ...,' since there is no property $F$, which depends only upon distances between pairs of points in $U$, such that $A$ has $F$ but $B$ does not. This may be clearer if we draw the universe $U$, as in figure 2.1. (Note that the border is not part of $U$, as $U$ consists only of $A$ and $B$.)

We single out $A$ from among the pair $\{A, B\}$ by means of the shapes of $A$ and $B$ themselves, as $fA = B$ and all distances are invariant under $f$ (which is a reflection in this diagram). Moreover, we cannot pick $A$ out from $\{A, B\}$ by means of its relationship to the rest of $U$ either, using only distances, since $f(A \cup B) = A \cup B$. For consider some relation $R$, definable in terms of distances, which $A$ bears to the rest of $U$. (We may think of this as $A$'s view of $U$.) Since $U$ as a whole is invariant under $f$, and $f$ maps $A$ to $B$, $B$'s view of $U$ is distance-indistinguishable from $A$'s view. Thus, $B$ must bear $R$ to the rest of $U$ as well, which means that $R$ cannot be used to single out $A$.

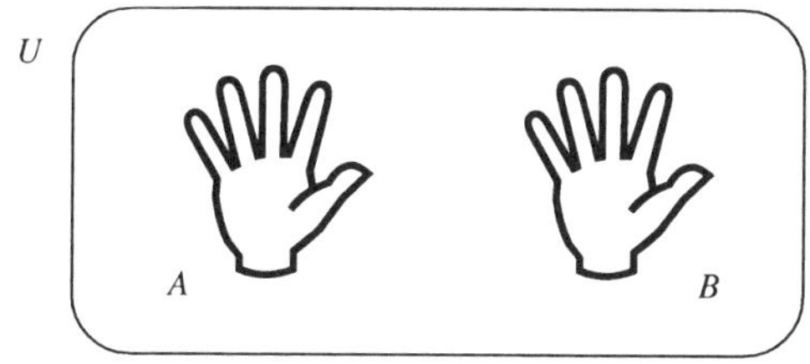


Figure 2.2

This matter of not being able to 'single out' either of *A* or *B* may become clearer if we consider another universe where *A* and *B* are not symmetric, such as the one in figure 2.2. Here there is an isometric transformation *f* such that $fA = B$, namely, a horizontal translation, but we do not have $fU = U$ since *f* maps *B* even further to the right rather than onto *A*. It is also clear that, while *A* and *B* are internally the same shape, their views of *U* differ. *A* sees another hand with the thumb pointing away from him, whereas *B* sees another hand with the thumb pointing towards him. The figure *A* can thus be picked out within *U* as the hand whose thumb is pointing at another hand in *U*.

This feature of a symmetric pair of objects, *A* and *B*, that neither can be uniquely singled out from among the pair $\{A, B\}$, is not a mere curiosity but the very property of symmetry that (in the context of propositions, rather than geometrical figures) gives rise to the symmetry axiom. Indeed, we may consider the fundamental definition of symmetry (at least the kind of symmetry required for the symmetry axiom) to be as follows: *A* and *B* are symmetric in *U* just in case neither can be singled out from among $\{A, B\}$ within *U*.

It is crucial to recognize the difference between symmetry and similarity. Symmetrical objects may be dissimilar. Consider, for instance, two right-handed gloves of the same type. The relation between these objects is similarity, or qualitative identity. There is no *contrast* between them. Now think of two gloves of the same type, but where one is left handed and the other right handed. They are not similar, as there is a contrast between them. This contrast emerges clearly if we imagine that someone is given a pile of left-handed and right-handed gloves, all mixed up. He will have no difficulty in separating them into two distinct piles, one of left-handed gloves and the other of right-handed ones, but he cannot tell which pile is which unless he has some previously identified right-handed (or left-handed) object (a right hand or a right-handed glove) to compare them to.

If two objects are similar then they are also symmetric, but the converse does not hold, as the glove example shows. Two objects that are not similar may still be symmetric, if the contrast between them is of a kind that does not allow either one to be uniquely singled out. Of course, the possibility or otherwise of using a dissimilarity between $A$ and $B$ to pick out (say) $B$ depends on the resources available.

Our definition of geometrical symmetry is then as follows.

**2.7.3 Definition** $A$ is symmetric to $B$, which we abbreviate to $A \Upsilon B$, within $U$ iff, for some isometric transformation $f$, $fU = U$ and $fA = B$.

**2.7.4 Theorem** Symmetry is a symmetric relation, that is, $A \Upsilon B \Rightarrow B \Upsilon A$ (all within $U$).

*Proof*: Suppose that $A \Upsilon B$, within $U$. Then, for some isometry $f$, $fU = U$ and $fA = B$. Now, since $f$ is isometric, so is $f^{-1}$. Then $f^{-1}U = f^{-1}fU = U$, and $f^{-1}B = f^{-1}fA = A$. Thus, $f^{-1}$ is the required isometry to make it the case that $B \Upsilon A$ within $U$.■

**2.7.5 Theorem** Symmetry is also reflexive and transitive, and thus an equivalence relation.

*Proof*: $A \Upsilon A$ within $U$, since the identity transformation is an isometry and maps $U$ to $U$ and $A$ to $A$. To show that $\Upsilon$ is transitive, suppose that $A \Upsilon B$ and $B \Upsilon C$, by virtue of isometries $f$ and $g$ respectively. Then the composition $gf$ is also an isometry that maps $U$ to $U$, and $gfA = gB = C$. Thus $A \Upsilon C$, as required.■

**2.7.6 Theorem** If $A \Upsilon B$ within $U$, then it is impossible to single out $A$ from among $\{A, B\}$ using properties that supervene on distances within $U$.

*Proof*: Suppose that $A \Upsilon B$ within $U$, by virtue of isometry $f$. Then, for the sake of reductio, assume that $A$ can be picked out within $U$ by means of some relation $H$ that holds between $A$ and $U$; that is, $H(A, U)$ holds. Since $H$ is invariant under $f$, it follows that $H(fA, fU)$ also holds. But then, since $fA = B$ and $fU = U$, it follows that $H(B, U)$ holds as well, so that $H$ does not in fact allow $A$ to be picked out from $\{A, B\}$.■

So far we have only considered $A$-$B$ symmetry within a universe $U$ consisting only of $A$ and $B$ themselves. This is an unnecessary restriction, as Definition 2.7.3 and the resulting theorems can be applied to any geometrical

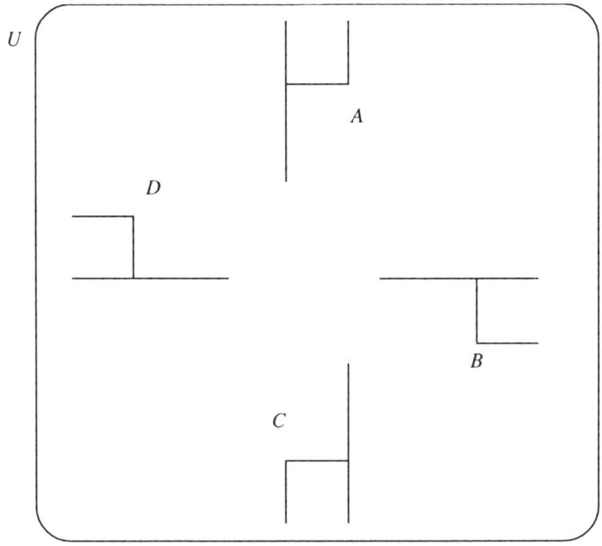


Figure 2.3

universe at all. Consider, for instance, the universe in figure 2.3. The universe $U$ now contains four figures, $A$, $B$, $C$ and $D$, and we see that $A$ and $B$ are symmetric in $U$. This is because there is an isometry $f$, namely, a rotation of $\pi/4$ about the obvious point, such that $fA = B$ and $fU = U$. It is also clear that it is impossible to single out $A$ from among $\{A, B\}$ within $U$ using only distances, as we should expect from Theorem 2.7.6. $A$ and $B$ are not the only symmetrical pair within $U$, as $A$ and $C$ are also symmetric for example. In fact, every pair of figures in $U$ is symmetric in $U$.

I have so far discussed the notion of relational symmetry in the context of geometry, as it is a simple way to convey the basic idea. We are really interested, however, in symmetry between *states of affairs*. Is it possible to apply the definition we have of '$A$ and $B$ are symmetric within $U$' to states of affairs $A$, $B$, and $U$? What would this intuitively mean? Let us consider the pure epistemic state $K_U$. We shall now consider two states of affairs, $A$ and $B$ say, within $K_U$, that is, the relative states of affairs $A/U$ and $B/U$. The intuitive idea is that the relation of $A$ and $B$ being symmetric within the epistemic state $K_U$ is analogous to the geometrical case defined above. $K_U$ does not contain sufficient resources to single out $A/U$ from among the pair $\{A/U, B/U\}$. We therefore have the following definition.

**2.7.8 Definition** *A* is symmetric to *B*, which we abbreviate as $A \Upsilon B$, within $K_U$ just in case $A/U$ cannot, within $K_U$, be singled out from among $\{A/U, B/U\}$.

To get a firmer grip on this definition, let us think about similarity and symmetry between states of affairs, with the help of an example. Consider a pair of dice, *a* and *b*, that are intrinsically similar, having the same size, shape, weight, colour, and so on. As any high-school student of probability knows, there are two ways to obtain a total score of 11 with two such dice: you can have a 6 on *a* and a 5 on *b*, or a 6 on *b* and a 5 on *a*. Even though these two states of affairs 'look the same,' as the dice are similar, they still count as distinct possibilities as the dice are distinct entities. The two states of affairs are not similar, as there is a contrast between them, but they are symmetric. The perfect similarity of the dice prevents either from being uniquely singled out from among $\{a, b\}$, and so neither of the states of affairs can be singled out either.

I can see no difference, with regard to states of affairs, between similarity and identity. Similarity between states of affairs seems to be sufficient for identity. If there is no contrast between states of affairs *A* and *B*, then they are the same state of affairs.

Now that we have a definition of symmetry, we are ready to state the Symmetry Axiom. It is as follows.

**Axiom 5** If $A \Upsilon B$ within $K_U$, then [\$1 if *A*] ~ [\$1 if *B*] within $K_U$.[23]

*Proof*: By symmetry(!) it is sufficient to prove that [\$1 if *A*] ≥ [\$1 if *B*], that is, that there is no objection within $K_U$ to giving up [\$1 if *B*] in return for [\$1 if *A*]. Now, assuming $A \Upsilon B$ within $K_U$, neither contract can be singled out within $K_U$. Thus, one cannot single out the case of possessing [\$1 if *A*], as compared with the case of possessing [\$1 if *B*]. (One can see that these cases are distinct, but one cannot pick out either case.) This being so, a rational person does not object to an exchange of [\$1 if *B*] for [\$1 if *A*], as required.■

**2.7.10 Theorem** If $A \Upsilon B$ within $K_U$, and *val*[\$1 if *A*] (within $K_U$) exists, then $Pr(A \mid U) = Pr(B \mid U)$.

*Proof*: By Axiom 5, if $A \Upsilon B$ within $K_U$, then [\$1 if *A*] ~ [\$1 if *B*] in $K_U$. If *val*[\$1 if *A*] (in $K_U$) also exists, then by Theorem 2.5.5 it follows that *val*[\$1 if *A*] = *val*[\$1 if *B*] in $K_U$, that is, $Pr(A \mid U) = Pr(B \mid U)$.■

**2.7.11 Theorem** If $A_1/U$, $A_2/U$, ..., $A_n/U$ are pairwise symmetric within $K_U$, and also pairwise mutually exclusive and jointly exhaustive within $K_U$, then $Pr(A_i \mid U) = Pr(A_j \mid U) = 1/n$.

*Proof*: Since $A_1/U$, $A_2/U$, ..., $A_n/U$ are mutually exclusive and jointly exhaustive within $K_U$, we have that {[\$1 if $A_1$], [\$1 if $A_2$], ..., [\$1 if $A_n$]} ~ \$1. Also, since they are pairwise symmetric, Axiom 5 entails that [\$1 if $A_1$] ~ [\$1 if $A_2$] ~ ... ~ [\$1 if $A_n$]. Then, using independence, it follows that {[\$1 if $A_i$]$_1$, [\$1 if $A_i$]$_2$, ..., [\$1 if $A_i$]$_n$} ~ \$1; that is, $n$ separate contracts [\$1 if $A_i$] are together worth \$1 as well. By Def. 2.3.4 we thus get *val*[\$1 if $A_i$] = $1/n$, and then by Theorem 2.3.5 we deduce that *val*[\$1 if $A_i$] = *val*[\$1 if $A_j$] = $1/n$. The result is immediate.■

This latest theorem bears more than a passing resemblance to Keynes's Principle of Indifference, and its predecessors. This should be a cause for concern since, while many issues in the foundations of probability are disputed, there is an impressive consensus that all such principles are invalid. Van Fraassen states, for instance, 'I regard it as clearly settled now that probability is not uniquely assignable on the basis of a Principle of Indifference, or any other logical grounds.'[24] Howson and Urbach concur: '... "logical" probability measures, whether based on the Principle of Indifference or on some other method of distributing probabilities a priori, do not, we believe, possess a genuinely logical status. For such systems are ultimately quite arbitrary, and we take logic to be essentially noncommittal on substantive matters.'[25]

Why are these authors so sure that any equi-probability principle must fail? There are two main reasons. First, there are the so-called paradoxes that these principles generate, whereby different ways of applying them to the same problem produce inconsistent results. Second, as mentioned in §2.1, there is the fact that purportedly a priori measures, such as Carnap's $c^*$ measure,[26] which are intended to provide a purely logical basis for inductive inferences, clearly contain synthetic assumptions about the world. These reasons do not seem to count against Theorem 2.7.11 in the least, however. It is true that this theorem is of no use at all in giving a theory of inductive inference but, as explained in §2.1, inductive inferences are not purely logical and so this is just as it should be. Moreover, as is demonstrated below by examining some examples, this definition of symmetry cannot be applied to a single problem in different ways, yielding inconsistent conclusions.

Let us look at some previous formulations of the Principle of Indifference, and the problems they face. According to Howson and Urbach, a typical formulation of the Principle from Bernoulli to Keynes is as follows:

> if there are n mutually exclusive possibilities $A_1, \ldots, A_n$, and *U* gives no more reason to believe any one of these more likely to be true than any other, then $Pr(A_i \mid U)$ is the same for all *i*.[27]

This principle is significantly more liberal than Theorem 2.7.11 in a manner that is highlighted by the following example. Let *U* be the state of affairs that a certain bag contains a nickel and a dime, one of which is drawn out, and let *A* be that the nickel is drawn, and *B* the dime. We should now consider the two contracts [$1 if *A*] and [$1 if *B*] from the point of view of *U*. According to the above principle, *Pr*(*A* | *U*) = *Pr*(*B* | *U*), since *U* gives us no reason to suppose that either *A* or *B* is more likely to be true that the other. From this it follows that [$1 if *A*] ~ [$1 if *B*], but does Axiom 5 give the same result? It does not, as it is easy to single out either state of affairs, and so *A* and *B* are not symmetric within *U*.

It seems very reasonable, in this example, to rule out the relations [$1 if *A*] > [$1 if *B*] and [$1 if *A*] < [$1 if *B*] between these contracts, since *U* does give us no reason for a definite preference of one over the other. Such a preference surely would be irrational, as it would be under-motivated. The principle seems then to conclude that [$1 if *A*] ~ [$1 if *B*], by elimination, but this is incorrect in my view since there is a fourth possibility, written [$1 if *A*] <> [$1 if *B*], that the contracts are incommensurable. Nothing of interest follows from the relation [$1 if *A*] <> [$1 if *B*]; e.g., it does not follow that *Pr*(*A* | *U*) = *Pr*(*B* | *U*). The above principle is therefore invalid, as lack of grounds for supposing the probabilities to be different is insufficient to make them equal.

A clearer example of two incommensurable contracts is as follows. *U* says that there are two urns, *a* and *b*, each of which contains only black balls and white balls. The proportions of black balls in each urn are unspecified, but *U* does say that the proportion in *a* lies in the interval [0.1, 0.5], and that in *b* lies in [0.2, 0.3]. A ball is selected from each urn. Let *A* be the state of affairs that the ball from *a* is black, and *B* that the ball from *b* is black. We are able to distinguish between *A* and *B* within $K_U$, as *U* gives different information about the two urns, so *A* and *B* are not symmetric. It also seems clear, however, that there is no reason to prefer the contract [$1 if *A*] over [$1 if *B*], or vice versa; these contracts are therefore incommensurable.[28]

We have shown that symmetry is an equivalence relation, which means that granting equi-probability on the basis of symmetry is safe from outright contradiction. The same is not true for giving equi-probability to any pair of propositions *A* and *B* such that [$1 if *A*] <> [$1 if *B*], since <> is not an equivalence relation. In particular, the relation is not transitive, as a slight addition to the urn example shows. Let *U* also say that urn *c* has a proportion of black balls in

[0.4, 0.45], and that a ball is selected from *c*, with *C* saying that it is black. Then there is no reason to prefer [$1 if *C*] over [$1 if *A*], and no reason to prefer [$1 if *A*] over [$1 if *B*], but [$1 if *C*] is definitely preferable to [$1 if *B*]. It is therefore inconsistent to assert $Pr(C \mid U) = Pr(A \mid U)$ and $Pr(A \mid U) = Pr(B \mid U)$, as it implies that $Pr(C \mid U) = Pr(B \mid U)$. Even in cases where there is insufficient reason to give different probabilities to a pair of states of affairs, one may not be warranted in giving them the same probability, for they may not have probabilities at all.[29]

We should not be surprised therefore if this Principle of Indifference leads to contradiction (which it does, as everyone knows). The most famous examples of such contradictions are the ones developed by Joseph Bertrand,[30] a selection of which we will now consider.

Consider a cube, whose edge length is given to be less than 2 cm. What is the probability, given this information, that the edge length is less than 1 cm? If we let the random variable **L** represent the edge length of the cube, then we see that the intervals (0,1] and (1,2] are of equal (Lebesgue) measure. Then, letting *A* represent $\mathbf{L} \in (0,1]$ and *B* represent $\mathbf{L} \in (1,2]$, we seemingly have no reason to regard *A* as more likely than *B*, or vice versa, so the classical Principle of Indifference declares them equi-probable. The problem arises when we realize that the information we have, that $\mathbf{L} \in (0,2]$, may equivalently be expressed as $\mathbf{L}^3 \in (0,8]$. $\mathbf{L}^3$, of course, represents the volume of the cube, another perfectly acceptable physical quantity. Assigning equal probability to sets of equal measure in (0,8], however, results in *B* being seven times as likely as *A*!

Another famous problem, known as Bertrand's Paradox, concerns the length of a chord. Given merely that *AB* is a chord on a given circle, what is the probability that the length of *AB* is greater than $\sqrt{3}$ times the radius of the circle? In other words, what is the probability that *AB* is longer than the side of an equilateral triangle inscribed in the circle? (Call this proposition *E*.) Bertrand provides three incompatible solutions to this problem, each one justified by the Principle of Indifference, which are as follows.

(a) Let us suppose we are given the location of *A*, one end point of the chord. (By the rotational symmetry of the problem, this should not affect the probability.) We then consider the angle between *AB* and the tangent to the circle at *A*. The angle lies in $[0, \pi]$, and *E* is true iff the angle is in $[\pi/3, 2\pi/3]$, so $Pr(E) = \frac{1}{3}$.

(b) Let us suppose we are given the direction of *AB*, so that *AB* is limited to a set of parallel chords. (By the rotational symmetry of the problem, this should not affect the probability.) There is a unique diameter to the circle that is perpendicular to these chords, so let us consider the point of intersection $P$ of $AB$ and this diameter. $P$ determines $AB$, and vice versa, but $P$ may lie anywhere on the diameter. It is easy to show that $E$ is true iff $P$ is within half a radius of the centre of the circle. Assigning equal probabilities to equal segments of the diameter, we deduce that $Pr(E) = \frac{1}{2}$.

(c) Excluding the case where $AB$ is a diameter, the position of $AB$ is uniquely determined by the position of its middle point $M$. E is true, of course, if $M$ is less than a half-radius from the centre of the circle. The area of this region is one quarter of the area of the circle, so $Pr(E) = \frac{1}{4}$.

These examples are devastating and unanswerable, showing decisively that the classical Principle of Indifference is invalid. There have been previous attempts to rehabilitate this principle, however, the best of which (in my view) being that of J.M. Keynes. Keynes's version bears many similarities to my own Axiom 5, but is firmly rejected by current scholarship. It should therefore be examined, so that its similarities and differences to my own approach may become clear.

Keynes sees the need for the Principle of Indifference to contain positive, rather than negative, criteria for equi-probability. The mere *absence* of grounds for assigning different probabilities is not enough; we require instead the *presence* of grounds for assigning equal probabilities. What positive criterion is offered? It is that 'our relevant evidence ... must be symmetrical with regard to the alternatives, and must be applicable to each in the same manner'; and elsewhere, 'If this relevant evidence is *of the same form* for both alternatives, then this Principle authorises a judgment of indifference.'[31]

This idea of symmetry, or 'sameness of form,' of the evidence between different possibilities is the mainstay of Keynes's idea, but there are two other, less attractive, features that require mention. First there is the notion that only the *relevant* evidence need be symmetric with regard to the alternatives. The idea seems to be that we take our evidence $U$, and the pass it through a kind of filter that removes all information that is irrelevant to the relative likelihood of $A$ and $B$, to get an epistemic state $U'$. $A$ and $B$ are then equally likely relative to $U$ just in case $U'$ is symmetric over $A$ and $B$. Thus, in general, a judgment of indifference consists of two stages: first we deem certain data irrelevant to the issue, and second we find that what remains is symmetric between the alternatives.

The filtering out of irrelevant information makes Keynes's principle more liberal than it would otherwise be, since the information excluded as irrelevant is often sufficient to break the symmetry. To see this, let us consider one of Keynes's own examples (1921: 53–4). Suppose we know that an urn contains two balls, one white and one black, one of which is drawn out. Is the selected ball equally likely to be white as black? Keynes acknowledges here that 'we know of *some* respects in which the alternatives differ,' which presumably entails that the evidence is not symmetric with regard to these alternatives, but claims that 'a knowledge of *these* differences is not relevant.' Thus, according to Keynes, since the colour difference is irrelevant, the two balls have the same probability of selection relative to this evidence.

It should be agreed, I think, that judgments of irrelevance are an important and valid part of calculating epistemic probabilities, but they do not seem to fall within the realm of pure logic. In my view, which I defend below, they involve the human faculty of le *bon sens*, and so rely upon some synthetic knowledge of the world. They are thus important for epistemic probability, but inadmissible in logical probability.

If we consider again the irrelevance of colour in the urn example, why is it that differences of colour can be ignored? If the balls were different sizes, would that be irrelevant? What if we had a ball and a cube instead of two balls? Is shape irrelevant also? What if one of the balls is sticky? I suspect that, at some point in this list, the differences would be considered relevant, so why is colour irrelevant? The only reason I can see is that balls are usually selected from urns by a blindfolded person, who clearly is unable to distinguish colours by touch alone. With the logical probability $Pr(A \mid U)$, however, we consider the epistemic state consisting only of knowledge of $U$. There is no additional 'background' information concerning the customary ways of drawing balls from urns. Moreover, if this is what Keynes has in mind,[32] then it seems that he is confusing two kinds of independence. We are here dealing with what is called *logical* independence, or irrelevance, whereas for a blindfolded selector the colour is *causally* irrelevant. Now, it is shown in chapter 5 that causal independence gives rise to independence in the physical-chance function, but this is very different from independence in the logical-probability function. Finally, it seems that to filter out 'irrelevant' information takes us back towards the classical Principle of Indifference – the very last thing Keynes wants to do. For the question of what counts as a *reason* to prefer $A$ over $B$ is rather similar to the question of what counts as relevant information in considering the symmetry between $A$ and $B$.

The second unattractive feature of Keynes's principle is that he adds an extra necessary condition for equi-probability, that the alternatives be *indivisible*. The extra condition is motivated by cases where it seems that the relevant information is symmetric over $A$ and $B$, but that it is also symmetric over $A_1$, $A_2$, and $B$, where $A = A_1 \vee A_2$ and $A_1 \Rightarrow \neg A_2$. If symmetry alone is sufficient for equi-probability, and $A$ and $B$ are exhaustive, then the symmetry of $A$ and $B$ yields $Pr(A \mid U) = \frac{1}{2}$, but the symmetry of $A_1$, $A_2$, and $B$ gives $Pr(A \mid U) = \frac{2}{3}$. The condition of indivisibility forbids the declaration of equi-probability in precisely this sort of case, where one or more of the symmetric alternatives is divisible into sub-alternatives that are symmetric with the rest of the original set.

If such cases are indeed possible, where sub-alternatives are symmetric with the rest of the original set, then clearly the symmetry condition alone will not suffice. To patch up the principle with additional conditions seems ill advised, however. For, if such examples exist then it follows that symmetry (in Keynes's sense, whatever that is exactly) is not an equivalence relation. In the case above we have $A_1$ symmetric to $B$, and $B$ symmetric to $A$, but $A_1$ is not symmetric to $A$, so that the relation is not transitive. Now this seems plainly wrong, particularly if we think of symmetry as 'sameness of form.' If $U$ has the same form for $A$ and $B$, and for $B$ and $C$, then it must also have the same form for $A$ and $C$, since identity is transitive. Any relation that can be thought of as 'sameness of $f$,' so that $a$ is related to $b$ just in case $fa = fb$, must be an equivalence relation. If it turns out that symmetry on some account is not an equivalence relation, then that account is wrong and needs to be corrected. To cover up the problem with ad hoc tinkering is unacceptable.

Keynes has very little to say about the relation of symmetry itself, and I myself cannot see how sub-alternatives might be symmetric with some members of the original set. It is provably impossible within Def. 2.7.8, as is shown by Thm. 2.7.5. Keynes's revised principle has been attacked on account of the arbitrary nature of its indivisibility condition,[33] as well as for its perceived failure to avoid the problems raised by Bertrand.

Since Keynes has almost nothing to say about the relation of symmetry itself, I am not sure whether or not his principle escapes the Bertrand paradoxes. From my point of view, however, the question is of little importance, as I have shown that there are important differences between Keynes's principle and my Axiom 5. The differences, in summary, are as follows.

1 Axiom 5 is based on a highly informative definition of symmetry (Defs. 2.7.3 and 2.7.8), whereas Keynes's principle involves no discussion of symmetry at all.
2 Axiom 5, unlike Keynes's principle, does not allow information in $U$ to be dismissed as irrelevant.
3 The symmetry of Def. 2.7.8 is shown to be an equivalence relation, which means that no condition of indivisibility is required.

What needs to be shown is that Axiom 5 does not give rise to paradox in the sets of circumstances concocted by Bertrand. To this I now turn.

First there was the example of the cube, for which we can consider either its edge-length $\mathbf{L}$ or its volume $\mathbf{L}^3$. Given that $\mathbf{L} \in (0,2]$, what is the probability of $A$, that $\mathbf{L} \in (0,1]$? The first question is: Is $A$ symmetric to $B$, where $B$ says that $\mathbf{L} \in (1,2]$? If we consider an arbitrary world that entails $A$, and also one that entails $B$, we find that the cube in the world for $A$ always fits inside that of $B$. It is therefore easy to pick out the state of affairs $A$ from among $\{A, B\}$. In a similar way, equal-length intervals in the range of $\mathbf{L}^3$ are not symmetric either. There is therefore no contradiction here.

Now let us look at Bertrand's paradox. Which, if any, of the three solutions is correct? It turns out that none of them is, although (b) is the correct solution to a very similar problem. Solution (a) involves partitioning an angle of $\pi$ into three equal sectors, but are these symmetric? The outer two, $[0, \pi/3]$ and $[2\pi/3, \pi]$, actually are symmetric, as becomes clear if we consider the reflection about the diameter that includes the chord-end $A$. The middle sector, $(\pi/3, 2\pi/3)$, is however not symmetric with either of the other two. It can be singled out, for instance, as the one in the middle. Thus the first argument fails.

Argument (b) considers a set of possible positions for $AB$ that are all parallel, and perpendicularly bisected by a diameter $D$. The position of $AB$ is therefore determined by its point $P$ of intersection with $D$. The argument assumed that equal-length segments of $D$ are equally likely to contain $P$, but are these states of affairs symmetric? Some pairs are, but most are not. For instance, if we represent $D$ by the interval $[-r, r]$, then $[-r, 0)$ and $(0, r]$ are indeed symmetric; but $[-r, -r/2]$ and $[0, r/2]$ are not, as the former is closer to the circumference, and can be singled out as such from among the pair. Roughly speaking, a segment near the circumference is not symmetric to one that is close to the centre. The second argument fails.

Argument (c) considers the mid-point $M$ of $AB$, which (except in the case where $AB$ is a diameter) determines the position of $AB$. It assumes that equal areas inside the circle are equally likely to contain $M$. It is true that some such pairs of regions are symmetric, but most are not on account of their differing distances to the circumference. This argument also fails.

One may feel intuitively that argument (b) is somehow better than the other two. (If so, this is likely to be encouraged by the fact that, if one drops straws from a great height onto a floor with a circle drawn on it, then about one-half of the chords produced are longer than the side of an inscribed equilateral triangle, in accordance with solution (b).) We can make sense of this intuition using the symmetry axiom.

In the physical experiment where the straws are dropped, there is presumably no causal interaction between the falling straws and the circle below; thus, the final positions of the straws are causally independent of the circle's position. As I mentioned before, there is a link between causal and probabilistic independence (demonstrated in chapter 5), so we may wonder what solution may be obtained if we assume that the circle's position is irrelevant to the position of the full line *AB*. Under this assumption we find that solutions (a) and (c) are still invalid, but something similar to (b) is valid. This will now be demonstrated.

The chord *AB* is part of an infinite line; the chord *AB* determines the line *AB*, and the line *AB* together with the circle *C* determine the line segment (i.e., chord) *AB*. Let us then consider the line *AB* for now. The state of affairs *U* does not say anything about *AB* except that it intersects *C*, but let us consider *U′*, which also specifies the direction of *AB*. Thus, within $K_{U'}$, all the possible positions for *AB* are parallel to one another. Let us define the *rectangle lm* as the subset of these parallel lines that lie between two of them, *l* and *m*. *U* does not specify the diameter of *C*, but let us suppose that *U′* gives the diameter as 5 cm. Then precisely one of the rectangles of width 5 cm contains *C*. Since we have no information about the position of *C*, any two propositions specifying different 5-cm rectangles in which *C* lies are symmetric. Also, for any rectangle *R* of width 5 cm, the various sub-rectangles of equal width inside it are symmetric within *U′*, as each may be mapped to the others under translation, which leaves the plane as a whole unchanged.

Now suppose we are told that a particular 5-cm rectangle, *R**, is the one that contains *C*. Its sub-rectangles now lose their symmetry with one another, as we can use *C* to pick out some over others. Now we apply our (extra-logical) assumption that the position of *C* is irrelevant to the position of the full line *AB*. This means that we cannot use *C* to single out any equal-width sub-rectangle in *R** from among a pair of them, so they retain their symmetry. Then states of affairs of the form 'the full line *AB* is in rectangle $a_i$,' where the $a_i$ are equal-width sub-rectangles of *R**, are symmetric within *U′*. Now, even though we do not know where *C* lies within *R**, the line *AB* still determines the *length* of the chord *AB*. The original reasoning of (b) can now be used to show that the length exceeds that of a side of an inscribed equilateral triangle with probability $\frac{1}{2}$.

We are not done yet, as we have only shown that $Pr(E \mid U') = \frac{1}{2}$, and so $Pr(E \mid U)$ is still unknown. *U′* contains two extra pieces of information over *U*, namely, the diameter of *C* and the direction of *AB*. It is clear from the argument above, however, that the particular values of these two data do not affect the result, that the probability is $\frac{1}{2}$. Then, since $K_{U'}$ is an authority for $K_U$, it follows that $Pr(E \mid U) = Pr(E \mid U')$, by the Authority of Logic principle.

If this were a book on logical probability, then it would be necessary to illustrate my theory of *Pr* with many more examples. In view of the modest aim of this chapter, however, which is merely to show that a revised notion of logical probability is viable, this would be excessive.

# 3

# Causation and Determination

The purpose of this chapter is to investigate the relations of causation and determination, and in particular to argue that a cause need not determine its effects. This work is needed as part of the theoretical foundation for the causal theory of chance, which will be stated in the next chapter. A complete analysis of determination is offered that is similar to standard accounts. No analysis of causation is given, however, as I hold that causation must be taken as primitive. Thus, my view is very much in the same vein as that of Anscombe (1971), and indeed this chapter may be seen as an attempt to develop her position more fully.

In addition to supporting the claim that causes need not determine their effects, this chapter develops many ideas that are essential to other parts of the book. Most important among these are the distinctions between logical relations and physical relations, and between maximality and completeness.

## 3.1 CAUSATION

### *3.1.1 Three Varieties of Causation*

There is strong evidence that the word *cause*, as used in common English, has more than one sense.[1] The evidence for this claim comes from consideration of certain kinds of examples, where our ordinary causal intuitions are pulled in different directions. Some of these examples are discussed below, but first I would like to present my aim in this section.

The causal theory of chance involves one particular kind of causation, which must be specified as precisely as possible. Unfortunately this kind of causation cannot be analysed, so there is no hope of specifying the relation in that manner. The best approach, I think, is to distinguish the required sense of causation from all the others, by consideration of examples. In this section, therefore, I shall endeavour to catalogue the different varieties of causation, that is, to follow in Aristotle's footsteps. Later in the chapter it will be argued that the kind of causation required for the causal theory of chance is the most basic of them, and that the others can be defined in terms of it.

Let us start with this basic kind of causation. It corresponds closely to Aristotle's notion of *efficient* cause, so let us call it that. The efficient cause is, according to Aristotle in the *Physics*,[2] the 'primary source of the change' (194b, 30). An efficient cause brings about, or produces, its effect. The cause is the *source* of the effect, so that the effect is derived from the cause. An efficient cause and its effect are two particular objects, or two concrete events. For example, the efficient cause of a given footprint is some particular foot – the one that made the footprint. Similarly, the efficient cause of a crater is often a meteor, and the sun and moon are the efficient causes of the tides.

Salmon (1984: 139) describes efficient causation as causal *production*, and gives the following examples. 'When we say that the blow of a hammer drives a nail, we mean that the impact produces penetration of the nail into the wood. When we say that a horse pulls a cart, we mean that the force exerted by the horse produces the motion of the cart. When we say that lightning ignites a forest, we mean that the electrical discharge produces a fire.' Like me, Salmon believes that causal production cannot be analysed in any of the ways proposed so far.

An efficient cause is always either partial or total. A total efficient cause is one that brings about its effect all by itself – it is solely responsible for the effect. A partial cause is merely a contributing factor in the production of an effect; it is merely part of the total cause. If a man pushes a boulder over the edge of a cliff all by himself, then he is the total cause of its falling. If he has help, that is, if others also exert force on the boulder, then he is merely a partial cause of its falling. The collection of *all* the people who helped push the boulder is a total cause of its falling. In a similar way, a child's mother is a partial cause of it, and the father is also a partial cause. Together they are a total cause, or something close to it (one should perhaps also count the food that the mother consumes during pregnancy as a partial cause, for example).

Total and partial (efficient) causation are both transitive relations. If $A$ is a partial cause of $B$, which is in turn a partial cause of $C$, then $A$ surely helped to bring $C$ about. $A$ is part of $C$'s causal history. For example, if I helped Fred write his CV, and this CV helped him to get a job, then I helped Fred to get a job. Also, if $A$ is a total cause of $B$, which is a total cause of $C$, then $A$ is a total cause of $C$. For an example of this, consider a collection of dominoes that are arranged upright, each on its edge, in a long line, on a smooth floor. The separation between each pair of dominoes is about half the length of each domino, so that when the first is pushed, this causes them all to fall down, one by one, in a chain reaction. In this example, the fall of each domino is a total cause of the fall of its successor, as nothing else exerts any force on that successor.[3] From this it follows that the fall of the first domino is a total cause of the fall of the last.

The chain of dominoes is a standard example of a *causal chain*, which is defined as follows. A causal chain is a sequence of discrete events, each of which is a cause (partial or total) of its successor. The continuous version of a causal chain (i.e., a causal 'rope,' if you will) is known as a causal process. A causal process may be (arbitrarily) divided into sections of finite duration, known as events, 'time slices,' or stages, where each such stage is either a partial or a total cause of its successors. Most causal processes are in continuous causal interaction with the rest of the world, so that each stage is only a partial cause of its immediate successor – the environment is also a partial cause. When a hammer drives a nail into a piece of wood, for example, we can say that the motion of the hammer is an efficient cause of the motion of the nail. Alternatively, we can talk in terms of a continuous causal process: the hammer begins to move, picks up speed, strikes the head of the nail, and the nail moves sharply downward. This process can be divided into as many events or stages as you please, where each stage causes its successor.

It seems to be the case that, if one event causes another (partially or totally), then those events are either (1) contiguous in spacetime, or (2) connected by a continuous causal process. For example, we usually think that, in order to make a footprint at a certain spot, at a certain time, the foot has to be in contact with that spot at that time. Also, while I can break a window without ever being within fifty feet of it, this does require throwing a stone or some such device; in other words, I have to connect the window to myself via a continuous causal process.

Salmon (1984: 141) has a nice example of a process that, while continuous, is not causal. Consider a beacon that delivers a tightly focused beam of light. If the beacon is pointed at a wall, then it will produce a bright, circular spot on the wall. As the beacon is rotated, the spot will move smoothly and continuously along the wall, and so the bright spot is a continuous process. The moving spot is not a *causal* process, however, since each stage of the spot is not caused by its predecessor, but rather each stage is caused by the beacon. If there is a gap in the wall, for example, then the spot process will have a discontinuity in it. The spot will vanish for a short period, and then reappear on the other side of the gap. This could not happen if the process were causal.

The word *cause* in English has other senses besides the one above. This is shown, for example, by cases of causation by omission. Suppose I am house-sitting for some friends, and one of my few responsibilities is to the water the plants. I forget to do so, and some of them die. I am therefore responsible for the death of those plants, and we can even say that my negligence *caused* the plants to die. It is clear, however, that I am not even a partial *efficient* cause of the death of the plants. I never went near those plants, and am not connected to them via any causal process,[4] so I did not cause them to die in that sense. They died of natural causes; that is, their water evaporated and they dried up. My actions do not appear in a complete causal account of their death.

When we say that I killed the plants, by not watering them, this seems to be motivated largely by the counterfactual belief that, had I watered the plants, they would not have died. Moreover, the fact that I was *supposed* to water the plants seems relevant, as of course no one else watered the plants either, and they did not thereby kill the plants.

Cases of 'causation by double prevention' also seem to involve counterfactual dependence. Suppose an ambulance, on its way to help Fred (a severely injured person), gets a flat tire. This causes the ambulance to arrive too late, after Fred is already dead. If the ambulance had arrived sooner, it would have brought him to hospital in time to save his life. Did the flat tire cause Fred to die? It seems acceptable, in English, to say that the flat tire was (at least partly) responsible for Fred's death, that is, that it caused him to die. This seems to be due to the counterfactual belief that, had the tire not gone flat, Fred would not have died. In the sense of *efficient* causation, however, Fred's death was caused entirely by his injuries; he was completely unaffected by the flat tire, which occurred some distance away, and which was not connected to him via any causal process.

We will return to the topic of counterfactual dependence, and its relation to efficient causation, below. Let us now consider another kind of causal claim, which always involves event *types*, rather than particular events. For example, we say that smoking causes cancer. This seems to mean that, for most people at least, smoking will increase that person's *chance* of getting cancer. A smoker is more likely to get certain kinds of cancer than a similar person who does not smoke.

Note that, in the sense of efficient causation, all kinds of things are causes of a particular cancer. If the person smoked, then that may be one cause, but also the food eaten, the water drunk, and the air inhaled are all efficient causes of the cancer. They are all part of the story of how the cancer came to exist. The material of the tumour came from food eaten, transported in the body's own blood, pumped to the tumour by the heart. The tumour is as dependent on the ordinary workings of the body as any 'legitimate' tissue is. It would be misleading to say, however, that drinking water is a cause of lung cancer, so there is clearly another sense of 'cause' here. Let's call this *generic causation.*

So there are at least three kinds of causation: efficient causation, counterfactual dependence, and generic causation. This is not intended to be a complete list; my purpose is rather to clarify the notion of efficient causation by contrasting it with its closest relatives. More distant relatives, such as final causes, do not concern us here.

### *3.1.2 Causation and Support*

To examine the relation between efficient causation and counterfactual dependence it is helpful to look at a simple example of causation, where we understand fairly clearly what is going on. One such example is where one object supports another, that is, prevents it from falling under gravity. The supporting object causes the supported object to remain in place.

A support C helps to prevent an object E from falling, under the force of gravity. It does so by exerting an *upward force* on E which, to a greater or lesser extent, balances the gravitational force. Thus, we shall say that C is a (partial) support of E just in case C exerts some upward force on E, however small it may be. Moreover, C is a *total* support of E if it is, by itself, sufficient to prevent E from falling. The force exerted on E by a total support is equal and opposite to the weight of E.[5] The distinction between partial and total supports corresponds, as should be clear, to the above distinction between partial and total causes. It is also clear that the relations of partial and total support are transitive.

Is there a counterfactual notion of support, corresponding to the counterfactual notion of causation? It seems that there is; namely, something like the following:

C supports* E iff (i) C is present, and
(ii) E does not fall, and
(iii) If C were not present, all else being as it is, E would fall.

In many cases where C supports E, C also supports* E. For example, suppose a painter is working at the top of a ladder, so that the ladder is supporting him. The ladder also supports* him since, if the ladder were not there, and everything else were exactly as it is, he would fall to the ground.

It is clear, however, that support and support* are not the same concept. There are cases of support that are not cases of support*, and vice versa. Suppose, for example, that the painter is working on the outside of my house, supported by a ladder. He has brought his dog with him, which is unfortunate, as it is large and boisterous. My job, therefore, is to hold onto the dog, to prevent it from charging about, and (very probably) knocking the ladder out from under the painter. In holding the dog, I am supporting* the painter, as the three conditions are satisfied. I do not support the painter, on the other hand, as I exert no force on him.

The situation can be modified so that the ladder supports the painter, without supporting* him. Suppose that, for security, the painter is tethered to the top of the house by a climbing rope. The rope is not quite taut, so that the painter's movements are not restricted by it; it has perhaps a few inches of slack. Thus, we shall suppose, the rope exerts no upward force on the painter, and therefore does not support him, even partially. (The ladder is still a total support of the painter.) In this situation, if the ladder were not present, the painter would not fall, as the rope would then snap taut, and *become* a support for the painter. Thus, while the ladder supports the painter, it does not support* him.

It is clear, from the examples discussed, that support and support* are different relations. I have not shown, of course, that *no* counterfactual analysis of support is possible. After all, it is likely that the definition of support* can be modified in some cunning way so that, in the cases mentioned, support and support* are coextensive. I do think that no counterfactual analysis of support can be successful, however, for the following simple reason: support is a *physical* relation between concrete objects, namely the exertion of force, whereas support* is a *logical* relation between possible states of affairs. The same point is true of the difference between efficient causation and counterfactual dependence. Efficient causation is a physical relation between concrete objects, whereas counterfactual dependence is a logical relation between propositions. The difference between physical and logical relations will be explained in the next section.

The similarity between efficient causation and support supplies us with a second metaphor for causation, to balance Aristotle's description of an efficient cause as a *source* of its effect. While Aristotle's metaphor is helpful, we can also think of an efficient cause as *supporting* its effect. We can call the efficient cause an *ontological* support, as it supports the real existence of the effect.

### *3.1.3 Logical Entities and Concrete Objects*

In distinguishing efficient causation from other relations it is necessary to distinguish between concrete objects and logical entities. The basic kind of logical entity is the proposition, which (as argued in the previous chapter) may be either objective or subjective. (I take the term 'proposition' to include both Fregean *Gedanken* and possible objective states of affairs.) Propositions of both kinds are easily distinguished from concrete objects, by the following three differences.

1 A proposition *A* can be an argument in a logical operation, such as negation, conjunction, or disjunction. There exist propositions $\neg A$, $A\&B$, $A \vee B$, for example, where *B* is another proposition. On the other hand, the concrete event C cannot be an argument of a logical operation. The closest thing to a logical operation on C would be something like mereological summation.
2 Propositions stand in logical relations to one another. One proposition can entail another, for example, and propositions may be consistent or inconsistent. Clearly, one concrete object cannot entail, or be inconsistent with, another. Concrete objects, such as concrete events, stand in very different relations of mereological inclusion and disjointness.
3 A proposition is an *intelligible* object, an object of *thought*, in the sense that it is a thing that can be believed. A proposition is defined (see Definition 2.2.2.6) as an expansion of an epistemic state. If one believes that the moon has no atmosphere, for example, then the *thing* that one believes is the possible state of affairs of the moon's having no atmosphere. (One does so by forming the thought that the moon has no atmosphere.) The moon itself, the concrete object, is not an object of thought in this sense. One cannot simply believe the moon – that would be nonsensical. In a similar way, a concrete *event* cannot be an object of belief, in this sense, either. The object of belief would have to be a possible state of affairs concerning the concrete event.

Propositions are the basic kind of logical entity, but there are others, most notably *properties*. Properties can be negated, conjoined, and so on. Properties can also be mutually consistent or inconsistent, one property can entail another, and so on. (The property of being a bachelor, for example, entails the property of being unmarried.) It seems that properties should be defined, in some way, from possible states of affairs. For example, to define the conjunction $F \,\&\, G$ of properties *F* and *G* we would say something like: The object *a* has the property $F \,\&\, G$ just in case $Fa \,\&\, Ga$. The exact details of this do not concern us; the point is that properties are derived, in some way, from propositions.

In a similar way, we must distinguish between concrete objects and *types* of object. A type, such as *fish*, *table*, and so on, is some sort of property, or collection of properties. A token object, such as a particular fish or a particular table, is a concrete object.

In addition to the three differences listed above, logical and physical objects also differ sharply with regard to real existence. Real existence is essential to concrete objects and concrete events, but not to propositions. There are non-actual states of affairs, but there are no non-existent objects. This is not to say that there is no significant difference between actual and non-actual states of affairs – on the contrary, I will argue later that this difference is crucial. But the *logical content* of a possible state of affairs is unaffected by its being actual or otherwise. Its being actual, for example, does not make it logically stronger, or weaker. It does not entail any more, or fewer, other possible states of affairs as a result of being actual.

The logical content of a possible state of affairs does not depend on whether it is actual. Thus, the logical relations between states of affairs, such as entailment and consistency, are also oblivious to the question of real existence. The relation of efficient causation, by contrast, can only hold of concrete events. An event is not caused unless it really happens, so that it has concrete existence. Also, a non-existent event cannot be a cause of anything else.

It is true that we sometimes speak of objects that do not exist, and of events that did not occur. In these cases, however, we are of course talking about *types* of object or event that are not instantiated. For example, in Vancouver there has been talk of building a third crossing of the Burrard Inlet. At present, the third crossing of the Burrard Inlet does not exist, which is to say that no bridge with certain properties exists. The idea of a non-existent *concrete* object is absurd, as concreteness is the same thing as real existence.

In a similar way, it is true that we talk hypothetically about causation. Regarding the third crossing of the Burrard Inlet, for example, someone might say, 'If the third crossing were built, then it would surely reduce traffic congestion in the downtown core.' This looks like a causal relation between two non-existent events, whereby a hypothetical bridge efficiently causes a hypothetical reduction in traffic. In fact, however, it is a *logical* relation between two possible states of affairs. From the one, the other is inferred. More will be said about this later. For now, I will merely point out that a logical relation is always used (or implied) in such sentences. You cannot say simply, 'If C occurred, it would *in fact* cause E.' It has to be 'If C occurred, it would *necessarily* cause E,' or 'If C occurred, it would *probably* cause E,' or something like that. You have to insert a logical relation for the sentence to make sense.

My claim that the logical content of a proposition is independent of its actuality, or real existence, is closely related to the familiar view that existence is not a property. Properties, as stated above, are logical entities, whereas the issue of real existence or concreteness falls outside the logical realm. The element of concreteness, in a physical object, is also similar to what has previously been called the *substance*, or substratum. Locke has pointed out that we have no clear conception of substance, so we are reduced to saying that substance is 'something, I know not what.'[6] This is quite correct, in my view. After all, the substance is precisely that aspect of a concrete object that lies outside the logical realm, and is not transparent to the understanding. The substance is the residue of a physical object, after all that is intelligible has been removed.

In my view, many puzzling phenomena are essentially tied to the concrete aspect of a physical object. Consciousness, for example, cannot exist in the abstract, but only in a concrete object; this is at least part of the reason why consciousness is so difficult to understand. A full consideration of this issue of concreteness would unfortunately take us beyond the scope of the book, however.

### *3.1.4 Efficient Causation and Time*

The relation between efficient causation and time is discussed in detail in §5.7, but it may be helpful to summarize my position here.

The system of mechanics developed in chapter 5 assumes that we can partition the actual, concrete history of a system into disjoint 'time slices,' each of finite duration.[7] Each such time slice, it is assumed, can be treated as a separate physical system, having its own dynamical properties and boundary condition. The actual history of each time slice is a 'chunk' of history that falls within a bounded interval of time $[t_1, t_2)$, and so is a concrete event. It is claimed in §5.7 that each time slice is a total cause of one of its immediate neighbours, so that the causal relation defines a linear ordering on the set of slices. Using the metaphor of support, we can picture the time slices as a vertical stack of books. Each book in the stack (apart from the bottom one) is supported by those below, and each (apart from the top one) supports those above. The book at the very bottom represents the earliest time slice,[8] which gives being to all the rest.

Although there is this close relation between causation and time in the cosmos, I view causal relations as essentially timeless. I consider it intelligible, for instance, to say that the cosmos itself has a cause, even though (since time is part of the cosmos) the cause and the effect could not be temporally related. Indeed, since I hold that time is just a physical dimension, like the spatial dimensions, even the causal relations between time slices are timeless. Each supports, timelessly, its successor in existence.

It is in virtue of the causal structure of the cosmos that the past is 'fixed,' while the future is 'open.' Since the past is causally upwind of the present, actions in the present have no causal influence on the past. The past is ontologically deeper than the present. Present events are instrumental in bringing about future events, however, so the future is open to manipulation.[9]

### *3.1.5 Conditionals*

In order to clarify further the difference between physical and logical relations, let us look at the meaning of conditionals, including counterfactual conditionals.

Let us start, for the sake of simplicity, with indicative conditionals. I adhere to the basic idea of Ramsey's analysis of these.[10] If $A \rightarrow B$ (if $A$ then $B$) holds in $K$, then $A$ must be consistent with $K$, that is, $K$ must not include $\neg A$. Thus, $K$ may be expanded by $A$, so that the state $A(K)$ exists.[11] We then have:

**Definition** $A \rightarrow B$ holds in $K$ just in case $B$ holds in $A(K)$.

In other words, $A \rightarrow B$ is believed in $K$ to the extent that $A$ entails $B$, within $K$.

Now let us consider counterfactual conditionals. For simplicity, we shall look at just one special case, where a claim is made about some dynamical system **X**, of the form 'If $A$ had occurred, then $B$ would have occurred.' The proposition $A$ is a false claim about the history of **X** at time $t$. What is $K$, the background epistemic state, here? Let us suppose that it contains two facts: the dynamical laws and the actual history of **X**.[12] Thus, $\neg A$ is believed with certainty in $K$, so that $K$ cannot be expanded by $A$.

Instead of an expansion we have to *revise* K by A, yielding the epistemic state $K^A$. The basic idea of a revision is some sort of 'minimal' change of $K$ to make $A$ true in $K$. Since $A$ is a claim about the history of **X** at time $t$, we first change the state of **X** at time $t$ so that $A$ occurs in $K^A$. We then have to make other changes to the history of the system, to make it consistent with the dynamical laws.[13] There are two kinds of such changes, namely, changes to the 'causes' of the state at time $t$, and changes to the 'effects' of the state at $t$. (The quotes indicate the fact that these are not really causes and effects, but inferred causes and effects.) The causes of the state at $t$ are, as stated in section 3.1.5, the states prior to $t$, and the effects are the states after $t$.

These two kinds of change to the history of **X** are marked, in English, by different tenses. If $B$ concerns times after $t$, then we say 'If $A$ had occurred, then $B$ would have occurred.' If $B$ concerns times prior to $t$, then we state the 'back-tracking' counterfactual 'If $A$ had occurred, then $B$ would have had to have occurred,' or more clearly, 'For $A$ to have occurred, $B$ would have had to have occurred.' The former kind of conditional, where the likely effects of the state at $t$ are inferred, is by far the more common, so we shall focus on it. We can use Lewis's notation $A \mathrel{\Box\!\!\rightarrow} B$ for it.

**Definition** $A \mathrel{\Box\!\!\rightarrow} B$ holds in $K$ iff $B$ is an entailed effect, within $K$, of the state minimally modified to make $A$ hold.

The truth of a counterfactual conditional (within the epistemic state $K$) depends therefore on there being an appropriate logical relation between two possible states of affairs. Note that such conditionals implicitly involve the notion of efficient causation, in that $B$ must be an inferred *effect* of the state at $t$. The English sentence effectively means, 'If $A$ had occurred, then $B$ would have occurred *as a result*.'

It is very important to realize that the events that are inferred to occur in the system, as results of the state at $t$, are not the inferred effects of $A$ *alone*, but rather of the (hypothetical) state of the *whole system* **X** at time $t$. This hypothetical state includes not just the occurrence of $A$, but many other things as well. Consider again the situation where I support* the painter, who is working at the top of a ladder, by restraining his dog, which would otherwise knock the ladder over. If we modify the state so that I release the dog, then we infer *from the whole situation* the effect that the painter would fall. We cannot infer the descent of the painter simply from the release of the dog.

The sentence 'If $A$ had occurred, then $B$ would have occurred' implicitly assumes that the system **X** is deterministic. It could be written, without change of meaning, as 'If $A$ had occurred, then *necessarily* $B$ would have occurred.' The claim is true, in other words, if the occurrence of $B$ is *entailed* by the modified state at time $t$. If the system **X** is stochastic, then $B$ cannot be entailed in this way, so there is no fact of the matter whether $B$ would have occurred, if $A$ had. In the stochastic case there is only a *degree* of entailment of $B$ by the state at $t$, within $K$. This is expressed by saying: 'If $A$ had occurred, then $B$ would *probably* have occurred.'

The point of this section is to show that the truth-makers of counterfactual conditionals are logical facts, that is, relations of entailment, and partial entailment, between possible states of affairs. The point is a subtle one, since the entailment in question is between states of affairs concerning causes and effects.

### *3.1.6 Efficient Causation Is Physical*

In this section I will argue that efficient causation is not a logical relation, between states of affairs, but a physical relation between concrete objects.

This does not mean, of course, that causation has nothing to do with logical entities such as states of affairs and properties. After all, concrete objects *have* properties, such as being square, heavy, yellow, and so on, and these are often causally relevant. Suppose an arrow hits a target, and pierces it. In this process, the speed and mass of the arrow, and the shape of the arrowhead, are causes of its penetrating the target. A cause produces its effect largely by virtue of the properties it possesses, so that logical entities are involved in efficient causation. Thus, my claim is really that causation is not a *purely* logical relation. Above all, causation is not any kind of inferential relation.

The basic reason for this view is that events that do not occur do not have causes, and they have no effects either. Causes and effects always have real existence, or concreteness. If causation were a purely logical relation, then there would not be such a requirement of real existence. A second argument is that efficient causation, like concreteness, seems to be opaque to the understanding. While we do have a concept of efficient causation, it is little more than 'something, I know not what.' If it were a logical relation, then we would have a successful analysis of it by now.

It is held by some that causation is a relation between facts. I am not against this view, as long as facts are understood to have real existence, or an element of concreteness. Every fact is a possible state of affairs, and so is a logical entity, but it is not *purely* logical. A fact is more than a possible state of affairs, just as an object is more than a bundle of properties. Consider, for example, the claim that the cat's being on the mat caused the mat to be warm. The first state of affairs does not cause the second by virtue of its logical content alone, but also by virtue of its being concrete. The concreteness of the first brings about the concreteness of the second. Concreteness is like an infectious disease, in that a thing needs to have it in order to give it to another. Efficient causation is the spread of concreteness.

## 3.2 DETERMINATION

### *3.2.1 The Standard Definition*

Lewis's definition of determination is as follows (1979: 37): $C$ determines $E$ just in case for every dynamically possible world in which $C$ occurs, $E$ also occurs. (A dynamically possible world is one that satisfies all physical laws.) This definition, modulo some irrelevant differences in the wording, is now fairly standard.[14] Let us suppose that the proposition $\mathscr{L}$ describes the laws of physics for the system in question. We now have that $C$ determines $E$ just in case every logically possible world that satisfies $\mathscr{L}$ and $C$ also satisfies $E$, that is, $\Box(\mathcal{L}\ \&\ C \rightarrow E)$. Then, using the fact that $A$ entails $B$, that is, $A \Rightarrow B$, just in case $\Box(A \rightarrow B)$, it follows that $C$ determines $E$ if and only if $C$ entails $E$ relative to $\mathcal{L}$, that is, $\mathcal{L} \Rightarrow (C \rightarrow E)$.

Determination is therefore a *logical* relation, being in fact the relation of entailment relative to $\mathcal{L}$. It was argued in chapter 2 that logical relations such as entailment are epistemic, that is, they concern the thoughts of an ideal reasoner. Determination is a matter of knowledge, of being able (in principle, for a sufficient intellect) to infer one state of affairs from another.

Since determination is a logical relation, its relata have to be logical entities – that is, propositions – and cannot be concrete objects. It is meaningless to assert that one physical object necessitates, or entails, another. Thus, since we are using 'C' and 'E' for physical objects that are the relata of the causal relation, we need to define some corresponding logical entities for the determination relation, which we shall call *m*(C) and *m*(E). What are these entities, however?

According to the theory developed in chapter 2, there are two kinds of proposition: subjective thoughts and objective states of affairs. Since the determination relation is held to be objective, and not dependent on human knowledge, its relata must be objective states of affairs. As a first approximation, we might say that *m*(C) is a state of affairs concerning C. This definition of *m*(C) will not do as it stands, however, since for any concrete event C there are many different states of affairs concerning C. Consider, for instance, some particular, physical event C of Smith stubbing a toe. The states of affairs of Smith stubbing one of his toes and of Smith stubbing a toe on his left foot both concern the event C. Both of them are actual states of affairs. Which, if either, is *m*(C)? The problem here is that some states of affairs concerning C are more detailed than others; we might say that some 'contain more information' than others.

The obvious way around this problem is to define *m*(C) as the *maximal* actual state of affairs concerning the physical event C, and nothing else.[15] In what sense is *m*(C) maximal, that is, what ordering relation is being used? The ordering relation is just logical entailment or necessitation, so that *m*(C) entails all the other actual states of affairs concerning C.[16] One may worry that there is, perhaps, no maximal state of affairs concerning C, but instead an infinite sequence of such states of affairs, each of which entails its predecessors. In this case, however, the terms of the sequence are pairwise consistent, so that the entire sequence has a conjunction. This infinite conjunction entails each member of the sequence, and thus is maximal over the sequence (it may be viewed as the limit of the sequence, if you will). We then have that C determines E just in case, relative to $\mathcal{L}$, $m(\mathrm{C}) \Rightarrow m(\mathrm{E})$.

### *3.2.2 The Generalized Lagrangian*

In the previous section I introduced the symbol $\mathcal{L}$, calling it a description of the laws of physics for the system in question. In this section I shall be a little more precise about what $\mathcal{L}$ is, exactly. I call $\mathcal{L}_{\mathbf{X}}$ the *generalized lagrangian* for the system **X**.

The concept of a generalized lagrangian is a somewhat Aristotelian one, as we may say that it represents the dynamical nature of the system. In the *Physics*,[17] Aristotle says that every thing that exists by nature has 'a principle of motion and of stationariness (in respect of place, or of growth and decrease, or by way of alteration).' An alteration of a system is according to nature if it proceeds from the system's innate principle of motion. If a motion has an external cause, being produced by a force acting on the system from outside, then the motion is compulsory or unnatural. The downward gravitation of a dense body, for example, was considered to be a natural motion, a change produced by the body's internal impulses, and not the result of any external force.

By the end of the seventeenth century, it seemed that this teleological approach to mechanics was infertile and bankrupt. In the light of Newton's achievements, the only change that could be considered natural was the motion of a particle in a straight line, at a constant speed. Since this rarely (if ever) occurs, the idea of a natural motion was of little value. Even gravitation was considered to be the result of an external force.

The forces that produce motion were considered to be governed by *laws*, such as Newton's law of gravitation, and the law of action and reaction. Moreover, the relations between forces and motions were also governed by laws, such as the second law of motion. In general we may say that the idea of the dynamical nature (as a regulator of motion) was replaced by the concept of a law. The idea that motions are regulated by laws persists in the philosophical community even to the present day.

This situation changed when Joseph Lagrange (1788) developed a system of mechanics, for deterministic systems, that is rather more Aristotelian. It will not be necessary to develop Lagrange's theory here, as the details are widely available,[18] but the rough idea is to analyse motions not in terms of forces, but rather in terms of exchanges between different forms of energy. A particle in free fall, for example, is steadily converting its gravitational potential energy into kinetic energy, in such a way that the overall difference between these two is minimized. Indeed, the backbone of Lagrangian mechanics is now considered to be the teleological principle that the actual trajectory of a system is always the one that makes the action integral take a stationary value. (This is now known as Hamilton's Principle.) The action integral is simply

$$S = \int_{t_1}^{t_2} L(x(t),v(t)).dt,$$

where $L(x,v)$ is the difference between the kinetic and potential energies of the system, and is known as the *lagrangian* of the system. The lagrangian depends, then, on the gravitational field that is the source of the potential. In this way, the gravitational field is viewed not as an external force that manipulates the particle, causing it to undergo unnatural motions, but rather as a part of the total 'system.' The total system acts according to its own nature, which is (roughly) to minimize the action.

It turns out that any isolated system, no matter how complex, has a lagrangian function, provided that all the forces involved are conservative. This lagrangian captures all the dynamically relevant information about the system, so that Hamilton's Principle is sufficient to determine the actual motion of the system, from its lagrangian, once a boundary condition is supplied. In short, the lagrangian of a system **X**, together with Hamilton's Principle, may be viewed as a representation of **X**'s dynamical nature. It is natural for a system to behave in such a way that its action is minimized.

If a system **X** has a dynamical nature, then we can define $\mathcal{L}_{\mathbf{X}}$, the generalized lagrangian for **X**, as the maximal state of affairs concerning **X**'s dynamical nature. The state of affairs $\mathcal{L}_{\mathbf{X}}$ can then be used to define the determination relation, as in the previous section. Determination is entailment relative to the generalized lagrangian. The generalized lagrangian is a replacement of, and an improvement upon, the concept of a physical law. It does all the same work, and more.

Are these two approaches to mechanics, using laws and using dynamical natures, really any different, however? It may seem that they are, at bottom, fully equivalent, by the following argument. In Lagrangian mechanics one typically does not make explicit use of Hamilton's Principle. Instead, to calculate the motion of the system, one uses Lagrange's equations of motion, which for a single particle are just

$$\frac{d}{dt}\left(\frac{\partial L}{\partial v}\right) - \frac{\partial L}{\partial x} = 0.$$

Now, since $\partial L/\partial v$ is the momentum of the particle, and $\partial L/\partial x$ is (minus) the gradient of the potential, that is, the applied force, this equation entails Newton's second law of motion. Thus, it may seem, the introduction of a 'dynamical nature' for the system is a rather superficial change. It may be an aid to calculation, but it makes no difference in theoretical terms.

This argument faces four objections. First, the Lagrangian formalism is more general than the Newtonian one. It is possible to write down a lagrangian for many systems that cannot be analysed using Newton's methods. This suggests that the concept of a dynamical nature is fruitful. Second, it should be admitted that the introduction of a dynamical nature does not show the concept of a law to be invalid. Rather, it enables us to obtain a much better understanding of what a law of nature is. Suppose some proposition $A$ truly describes the actual history of a system **X**. Is $A$ a law, or merely an accidental fact? Using the notion of a dynamical nature for **X**, we can say that $A$ is a law (i.e., nomically necessary) just in case $\mathscr{L}_{\mathbf{X}} \Rightarrow A$. The point is that the idea of a dynamical nature is more fundamental and general than the idea of a law.

On this view, a law is characteristic of the dynamical nature of the system. Its necessity consists of the fact that it is *entailed* by $\mathscr{L}_{\mathbf{X}}$, which maximally describes that dynamical nature. The modal force of the law, in other words, derives from its logical relation to the dynamical nature. This account of laws solves the *inference problem*, as van Fraassen (1989: 29, 38–9) calls it, of showing that the inference of '$A$ is true' from '$A$ is a law' is valid. The inference is simply modus ponens; $A$ is inferred from $\mathscr{L}_{\mathbf{X}}$ and $\mathscr{L}_{\mathbf{X}} \Rightarrow A$.[19]

The third advantage in introducing the generalized lagrangian $\mathscr{L}_{\mathbf{X}}$ is that it offers a superior understanding of the dynamics of a stochastic system. This matter is discussed in detail in chapters 4 and 5, but the rough idea is as follows. Stochastic laws involve probability, which is a kind of modality. Now, if our fundamental account of the dynamics of a system is in terms of a stochastic law, such as a stochastic equation of motion, then the probabilities are built into the fundamental dynamical description of the system, and so their nature remains mysterious. The description $\mathscr{L}_{\mathbf{X}}$ of the dynamical nature, by contrast, does *not* involve probabilities. $\mathscr{L}_{\mathbf{X}}$ talks about what is there, in the system, rather than about what will probably occur. The probabilities arise from the logical relation of partial entailment between $\mathscr{L}_{\mathbf{X}}$ and propositions about the motion of **X**.

A stochastic system does not have a lagrangian in the usual sense, because the forces are random, and so the potential energy is not well defined. It does not follow, however, that the more general concept of a dynamical nature is not valid for stochastic systems. One of the main theoretical claims of this book is that *every* system has a generalized lagrangian.

The fourth advantage of the dynamical nature is that it can be understood as a *cause* of the motion of the system. Suppose a pendulum vibrates with a particular frequency. What is the cause of this frequency? It is easy to show that the frequency of a pendulum is approximately independent of both the amplitude of its motion and the mass of the bob. It depends only upon the length of the string. It seems clear, therefore, that the frequency of the vibration is brought about, or *caused*, by the length of the string. The length of the string is part of its intrinsic dynamical nature, represented in the generalized lagrangian, so that in general we can say that a system's dynamical nature is one of the causes of its behaviour. A law, by contrast, does not cause anything. It is merely a statement about the motion of the system.

I say that $\mathcal{L}_{\mathbf{X}}$ describes only *one* of the causes of **X**'s behaviour since, in the pendulum example, the dynamical nature is not sufficient to cause a particular amplitude of vibration. There can be two pendula that are exactly similar, so that they have the same generalized lagrangian, yet whose amplitudes of oscillation are quite different. This difference is due to the fact that one is initially given a large kick to begin its motion, and the other only a little push. Thus, a second cause of a system's behaviour is the manner in which it is set going, which is usually represented by a *boundary condition*.

### *3.2.3 Determinism*

Determinism has been defined as the claim that every event[20] has a cause. Those advancing this definition, however, believed that causation and determination are the same relation, or at least necessarily equivalent, a claim that I reject. If the relations are distinct, then it makes sense for 'determinism' to be defined in terms of the determination relation, rather than (or in addition to) the causal relation. So the rough idea of determinism is that every event is determined. But determined by what?

The usual answer is that, in a deterministic system, every event is determined to occur by past events, or the total past, in the system in question. Why is this, however? Why be interested in determination by past events, rather than future events (or a bit of both)? The answer is not that future events do not determine past events, for examples of this are very common. Consider the weather, for instance. It is doubtful whether November's weather is determined by the state of the world in October; at least, humans are not very good at drawing correct inferences about November's weather from information about October. By contrast, November's weather is very precisely determined by the state of the world in December, as in December there exist very precise records of November's weather, from which facts about the weather itself may be inferred with a high degree of certainty. We can 'predict' temperatures, pressures, levels of rainfall, and so on, with great accuracy.

The reason why one is interested in whether events are predetermined, I believe, is that causes temporally precede their effects. We are really interested in whether an event has causes that determine its occurrence. The causes of an event are contained in its past, so if an event is determined by its causes, then it is predetermined. We therefore have the following definitions.

**3.2.3.1 Definition** The *Principle of Universal Causation* is the claim that every physical event has a total efficient cause. There are no spontaneous events, or 'events from nowhere.'

**3.2.3.2 Definition** The *Determined Effect Principle* (DEP) is the claim that, if C is a total efficient cause of E, then *m*(C) must also determine *m*(E). In short, causes always determine their effects.

**3.2.3.3 Definition** *Determinism* is the conjunction of the Principle of Universal Causation together with the Determined Effect Principle.

## 3.3 HOW ARE CAUSATION AND DETERMINATION RELATED?

It was argued in §3.1.6 that causation is a relation between concrete events, whereas determination is a relation between possible states of affairs, so that the two relations are distinct. For the causal theory of chance, however, I need a stronger result, that a total efficient cause need not determine its effects, so that the Determined Effect Principle is false. In this section, therefore, I shall examine the relation between causation and determination.

### *3.3.1 Necessary and Sufficient Conditions*

Sometimes causation is explicated in terms of necessary and sufficient conditions, for example, that a cause is a necessary condition for its effect, or a sufficient condition for its effect, or both. These accounts may be seen as attempts to define causation in terms of determination, as C is sufficient for E just in case C determines E, and C is necessary for E just in case E determines C. It is clear, however, that *efficient* causation cannot be analysed in this way, as it is not a logical relation of any kind.

We must be wary of the word 'sufficient' in relation to causes. Suppose C is a total cause of E, so that C alone 'exerts ontological force' on E. Then, since E does in fact exist, it seems reasonable to say that C is sufficient for E. After all C, by itself, was *enough* to bring E about. This is purely a fact about what did happen, in the actual world. In this sense of 'sufficient,' one might reasonably say that a team's skill was in fact sufficient to win the game, but it might not have been. This notion of sufficiency should not be confused with the logical claim that C is sufficient to *determine* E, so that $m(C)$ entails $m(E)$ relative to $\mathscr{L}$. The latter says that *any* possible world, with our lagrangian, that includes $m(C)$ also includes $m(E)$. We must distinguish therefore between causal and determinative sufficiency. The ambiguous term 'sufficient cause' is to be avoided.

### *3.3.2 Some Examples*

The easiest way to show that causation does not imply determination, and vice versa, would be to provide examples where one of the relations clearly holds and the other clearly does not. In this section I shall examine such examples.

It is fairly easy to show that determination does not imply causation. The simplest argument for this is the fact that later events can determine earlier ones, whereas efficient causation is always from the past to the future. For example, present observations of the solar system determine the times and places of solar eclipses in the distant past, thousands of years ago, to a high degree of accuracy. From knowledge of the present state, and Newton's laws, it is possible in principle (and even in practice, in this case) to infer much earlier states with a high degree of accuracy and reliability.

One might respond to this by saying that determination plus the right temporal ordering implies causation, that is, if $m(C)$ determines $m(E)$, and C is temporally prior to E, then C caused E. Even this weaker claim, however, is vulnerable to the following sort of example.

Suppose Fred is in a room with two exit doors, A and B. Fred believes that he may leave the room through either door, but I have secretly locked door A, making it effectively impassable. At some point Fred decides to leave the room, and happens to choose door B for this purpose. (He did not try to open door A.) It seems clear that my locking door A was not an efficient cause of Fred's leaving through door B. It was not even a partial cause, as my action had no effect on Fred at all. However, given that door A is locked, it follows that Fred must leave through door B, so that my action does *determine* Fred's exit door.

To be more precise, it is the locking of the door in *conjunction with the other facts of the situation* that determines Fred's exit through door B. The information that door B exists, and is not locked, and that no other door exists, that Fred desires to leave the room, and so on, is all needed. The *whole situa-*

*tion* causes Fred's action, but the locking part does not. The locking (in conjunction with the rest) determines the action.

It is much more difficult to show, from examples, that DEP is false, that is, that efficient causation does not imply determination. Such an example must involve a system that is clearly indeterministic, so we have to take a quantum-mechanical system.[21] Such systems are not very well understood, however, so it is not easy to establish clear causal connections between events in these systems.

By far the strongest example of this kind is the EPR experiment, which, I believe, very clearly contains events that are fully caused, but not determined by their causes. A proper discussion of this experiment requires some length, however, and so is left until chapter 6. For now, we will look at a much weaker example, which Anscombe (1971: 24) discusses, namely, Feynman's bomb.

A bomb is connected to a Geiger counter, in such a way that if the registered intensity of radiation exceeds a certain threshold, then the bomb is detonated. Initially, since the required level of radiation greatly exceeds the mean background radiation, the chance of the bomb exploding (any time soon) is very low. A mischievous person then throws a chunk of radioactive material, which lands close to the detector, and the bomb immediately explodes. For convenience, let us call the act of throwing the chunk of material 'C,' and call the explosion event 'E.'

It seems possible that, in this scenario, the following claims both hold:

1 the ß-particles (say) that triggered the explosion *came from* the chunk of radioactive material, and
2 the throwing of the radioactive material did not *determine* that an explosion would occur at that time.

We know that ß-particles are produced by unstable nuclei, of which the chunk contains a great number, so it could well be the source of the radiation. This claim is consistent with claim 2, assuming that the standard view of radioactive decay is correct. (On this view, the time at which an unstable nucleus decays is not determined by its initial state.) Now, claim 1 seems to entail that C caused E, whereas claim 2 says that *m*(C) does not determine *m*(E).

This example, though it may persuade some, is not fully convincing, since it does not seem possible to establish the existence of a continuous causal process, from the throwing of the radioactive material to the explosion. In particular, it may be doubted that the decay of each unstable nucleus is fully caused by prior events. Where does the 'decision' to decay, at a particular time, come from? Two atoms may be exactly similar at one time, and yet a moment later one has decayed and the other has not. It may be claimed that this 'decision' came from nowhere; it is an uncaused event, that is, one that does not arise within a continuous causal process. The formalism of quantum mechanics certainly does not show us any mechanism that brings about individual decays. It does not explain why one atom decays, at a given time, while an exactly similar one does not.

### *3.3.3 Explanations and Mechanisms*

The claim that the decays of individual atoms must be uncaused seems to be founded upon certain intuitions about causation, such as:

1 If an event has a cause, then it can be explained.
2 Every causal process has some mechanism by which it operates.
3 Similar causes have similar effects. Thus, if exactly similar systems suddenly start to behave differently, then the behaviour of at least one of them is uncaused.

Nuclear decays do not seem to have explanations, at least not *full* explanations. If one asks why this nucleus decayed, a physicist may reply that it was unstable, and justify this claim by referring to the quantum mechanics of the nucleus, the strong nuclear force, tunnelling, and so forth. This is very well, as far as it goes, but one may ask a different question: 'Why did this atom decay, while that (similar) one did not?' The physicist has no reply to this, except to say that the processes are stochastic, so that similar experiments may have different outcomes.

Nuclear decays, and random events in general, seem to lack causal mechanisms. The existence of a causal mechanism, in the case of a random event, seems to be ruled out by the fact that an exactly similar system might have behaved differently. The exact similarity entails that any mechanism that existed in the one would have to exist in the other. In that case, of course, the similar mechanisms should have produced similar results.

This brings us to the third intuition, that exactly similar systems, which are in the same state, cannot ever diverge in behaviour. Or, if they did, then this would involve an uncaused event. This idea, which can be expressed as the claim that symmetry is always conserved in a causal process, is really the most fundamental of the three. It is developed, and refuted, in the next section.

### *3.3.4 The Symmetry Conservation Argument*

The basic idea of this argument, which I call the symmetry conservation argument,[22] is that an undetermined effect would permit the spontaneous emergence of asymmetry, which is impossible. Since a symmetrical cause must have a symmetrical effect, it follows that causes must determine their effects. The main premiss of the argument, that symmetrical causes have symmetrical effects, I call *conservation of symmetry*, which is defined as follows.

**3.3.4.1 Definition** *Conservation of Symmetry* (CS) is the claim that, if C is a total efficient cause of E, then (necessarily) any type of exact symmetry possessed by C is also possessed by E.

The idea that symmetry is conserved is well known, from the familiar story of Buridan's ass. The hungry donkey is faced with two tasty-looking bales of hay, but unfortunately the entire system[23] is exactly symmetric with respect to the two bales. It may be, for example, that there is some plane about which the system has perfect reflective symmetry. Each hay bale is the exact reflection of the other in this plane, and the beast itself is perfectly symmetric about the plane as well. If the ass were to move even an inch toward one of the bales, then the symmetry would be broken. Thus, the poor creature is doomed to starve to death, even though ample nourishment is readily available. In order for the donkey to eat either bale, the system would have to evolve into a state in which the reflective symmetry is broken, and this is impossible.

It is important to realize that CS only applies to exact, or perfect, symmetry. There is no guarantee that slight asymmetries will, under causal evolution, remain slight. It is well known, for example, that non-linear dynamical systems are highly sensitive to slight variations in the initial state, so that slight asymmetries can quickly become large. If the symmetry is not quite exact, therefore, there is no guarantee that it will be conserved as the system evolves.

The motivation for CS comes from the Principle of Universal Causation. The idea is that a symmetry-breaking event could not causally emerge from a symmetrical situation. The very first asymmetric state would have to be uncaused – an event from nowhere. Why would the first asymmetric event have to be uncaused, however? After all, the appearance of a red event within an initially non-red system need not be spontaneous, and the same is true for the first noisy event within an initially quiet system, and so on.[24] Why is the appearance of an asymmetry any different from the appearance of redness or a loud bang?

The reason for this becomes clear when we consider what it means for a system to be symmetric. The claim that the donkey system has reflective symmetry at time $t_0$ about the plane $\Pi$ is that, when the system is reflected about $\Pi$, its state remains unchanged, in some sense. The state of affairs after the reflection is no different from the state of affairs before the reflection. It follows from this that neither bale has any property (within the system)[25] that enables it to be singled out from among the pair of bales, for any property possessed by one bale would, under reflection, be possessed by the other bale. It seems to follow from this that the system has no power to select either bale over the other. How could it? Such a power, if it existed, would break the assumed symmetry between the bales within the system. Suppose, however, that the system were to evolve in such a way that, at time $t_1$, the ass is eating one of the bales. Since the state at $t_1$ is caused by the state at $t_0$, in this case the state at $t_0$ would have succeeded in selecting one of the bales (from among the pair) after all, which is impossible. In general, a symmetric state cannot have the power to produce, or generate, an asymmetric state.

Let us now proceed with the symmetry-conservation argument, that CS entails the determined-effect principle. The argument begins with the assumption that a system **X** is stochastic, so that its later states are not determined by its earlier ones. It is then shown that such a system violates CS. Therefore, if CS holds, it follows that **X** must be deterministic.

The argument must apply to all systems **X** for all initial states, so it will not do to assume that **X** has any kind of symmetry in itself. How then can we make use of the principle CS, therefore, in reasoning about **X**? The trick is to use *two* systems, **X** together with an exact copy of **X**, called **X′**. **X** and **X′** are causally isolated from each other (as well as the rest of the world), and they are exactly similar in the sense that they have the same generalized lagrangian. Moreover, suppose that at $t_0$ they are in the same state. Although neither **X** nor **X′** need have any symmetry in itself, the *combined system* ⟨**X**, **X′**⟩, the aggregate of the two systems, is symmetric at $t_0$. At $t_0$, the system ⟨**X**, **X′**⟩ has no power to select either **X** or **X′** over the other.

The system **X** is stochastic. Thus, even given a fixed state for **X** at $t_0$, there is more than one possible state for **X** at the later time $t_1$. The same is true for **X′**, since the two systems are dynamically similar. It is therefore possible that, at $t_1$, the two systems **X** and **X′** are in different states. If this were to occur, then the state of ⟨**X**, **X′**⟩ at $t_1$ would no longer be symmetric in the manner of its symmetry at $t_0$. A symmetric state would have evolved into an asymmetric one. We have thus shown that, under the assumption that **X** is stochastic, it is at least *possible* for symmetry not to be conserved. Now, since CS holds that symmetry is *necessarily* conserved, we have a contradiction with CS.

### *3.3.5 The Refutation of CS*

What is the flaw in the symmetry-conservation argument? The short answer is that the principle CS is deeply flawed. CS is not false, exactly, but rather based on a false presupposition, so that it lacks a truth value. The easiest way to see the problem with CS is to consider the case of two exactly similar systems, such as two radioactive atoms. What does it mean to say that the atoms are exactly similar? This term usually means that the atoms have exactly the same intrinsic properties, that is, all the same properties except spatial location. These atoms are concrete objects, however, so each is more than a collection of properties. Their properties may be the same, but that is not sufficient to make the objects exactly similar.

For an analogy, consider a pair of identical twins. They share the same genetic code, but that is not sufficient to make them exactly similar. After all, there is a lot more to a real person than his genetic code. In a similar way, there is a lot more to a concrete event C than the state of affairs *m*(C).

To talk of atoms as exactly similar to one another is to imagine them as abstract objects, like congruent solid figures, rather than as concrete things. This way of thinking is easy to slip into, as we cannot see atoms, and are only acquainted with them through abstract mathematical models.

If two concrete systems cannot be exactly similar, then it is meaningless to say what must happen to a pair of such systems. Thus, CS has no truth value. It will nonetheless be helpful to look at the argument used to support CS, and expose its flaws.

The argument for CS concerns the possibility of the system ⟨**X**, **X′**⟩ being able to select either **X**, or **X′**, from among the pair. Since **X** and **X′** share all the same properties, it seems that no such selection can occur. But, if the systems start to behave differently, for instance, if one atom decays and the other does not, then it looks like a selection has taken place. The flaw in the argument is that it conflates logical with physical selection. It is true that, within a maximal epistemic state concerning ⟨**X**, **X′**⟩, neither atom can be selected *in thought* from among the pair. It is not possible to 'point,' mentally, to either atom, as they are indistinguishable. But each atom is more than a set of properties, and causal processes are not logical entities. The causal processes that result in the 'selection' of one atom, when it decays, simply do not exist in the logical realm. They do not appear in the maximal epistemic state concerning the systems. Thus, the argument equivocates on the word 'selection.'

The impossibility of distinguishing, within a maximal epistemic state, the events (**X** decays, **X′** does not) and (**X′** decays, **X** does not) does not guarantee that neither event can occur. It merely guarantees that the two events have the same probability, within the epistemic state, of occurring.

We now see that intuitions that call for full explanations of all events, and a mechanism for every causal process, are similarly mistaken. These intuitions also conflate logical entities with physical ones, by holding that we can comprehend, in principle at least, every causal process. But a causal process is a physical, concrete, thing, and so is not an object of thought (like a state of affairs). A causal process is, by its very nature, largely inscrutable to the understanding. (It is not *entirely* inscrutable since, as discussed above, physical objects do have properties, and these are causally relevant, so a partial understanding of a causal process is possible.)[26]

Let us think about the difference between a deterministic and a stochastic causal process. In a deterministic process, the states of affairs concerning the earlier events in the system fix the later states of affairs. Thus, the concrete aspect of the system plays no role in choosing what takes place in its history. As the process develops, each predetermined state of affairs merely becomes 'clothed' with concreteness. The concretizing of the history is therefore a mere epiphenomenon, and the causal process is transparent to the intellect. In a deterministic causal process, explanations and mechanisms are always available, at least in principle.

In a stochastic process, by contrast, the earlier states of affairs do not determine the later ones. Of course, a single history does emerge, as the process develops, so this history is created by the concrete aspect of the system, as well by as the earlier states of affairs. In a stochastic process, the logical and concrete elements are not cleanly separated, but deeply entwined. The properties of the system are causally relevant, but the concrete aspect is needed to choose a unique history from among all the possible histories (i.e., from all the histories that are consistent with the generalized lagrangian and the initial state).

### *3.3.6 Maximality versus Completeness*

In order to clarify my claim that exact similarity between two concrete systems is impossible, it may help to explain the difference between *maximal* similarity and *complete* similarity. Two physical systems may be maximally similar, but not completely similar. This distinction between maximality and completeness will also be useful later in the book, where we consider whether or not quantum mechanics is complete. My view is that quantum mechanics is more or less maximal, but it is not complete.

The notion of a maximum is derived from an ordering relation $\geq$ on a class $\Phi$, say. A member $x$ of $\Phi$ is a maximum of $\Phi$ with respect to $\geq$ just in case, for every y in $\Phi$, $x \geq y$. Now, to say that some object is maximal means that it is the maximum of the class of all possible objects of its kind. Thus, 'maximal' means (roughly) 'greatest possible.' Note that the concept of maximality depends essentially on an ordering relation – without any ordering, the concept is meaningless. For example, when $m(\mathbf{X})$ was defined as the maximal state of affairs concerning **X**, the ordering relation was logical entailment.

The notion of completeness, or perfection, is less clear and much harder to define. The rough idea of a perfect, complete (or finished) object is one that has arrived at some goal. In many cases it is a matter of matching, or conforming to, some external standard. The concept of completeness does not, therefore, require an ordering. Also note that the concept of maximality does not require an external standard.

To illustrate the difference between completeness and maximality, let us consider some examples. First let us think about long jumps. One can speak of the best long jump for a particular athlete, or the best jump ever achieved by a human being, and perhaps even the longest possible jump for a human being. Thus, the notion of a maximal (longest) jump makes some sense. The notion of a complete, or perfect, long jump seems to be meaningless, however. What is supposed to define the complete jump? There is no external standard of comparison here. Of course, if it is a matter of jumping across a gorge of fixed width, then a perfect jump is simply one that carries the jumper safely across. Such a perfect jump need not be maximal in length, however.

For another athletic example, think of the position of a runner in a race. The leader of the race, at a given time, has a position that is maximal though not complete (finished). Her position is ahead of all the other runners, but she has not yet reached the finishing line.

The laws of thermodynamics may dictate a maximum efficiency of 35 per cent, perhaps, for a particular heat engine. This falls short of complete (100%) efficiency, where all the thermal energy is converted to mechanical energy, but is nonetheless maximal, being the best efficiency possible. It is invalid to argue that, since this steam turbine is not completely efficient, it is possible to make it more efficient. This reasoning confuses incompleteness with sub-maximality. In a similar way, an engine might deliver its maximum power at 4800 rpm. Thus, at 4800 rpm the power is maximal, but it makes no sense to say that the power is *complete* at this angular speed.

We see that in some of these cases, but not others, completeness implies maximality. A perfectly efficient engine would also be maximally efficient, for example. There are also cases where maximality implies completeness. A glass that is maximally full of water, for example (so that it cannot be filled further) is also completely full (filled to the brim).

Why do I talk of $m(\mathrm{C})$ being the maximal state of affairs concerning C, rather than the complete state of affairs? First, it is defined as the strongest state of affairs concerning C, using logical entailment as the ordering relation.

Thus, the term 'maximal' is certainly appropriate. We have seen that, in some cases, a maximal object is also complete. So perhaps *m*(C) is also the complete state of affairs concerning C.

What would this mean? It would suggest that *m*(C) is just like C, that there is a perfect match between them. We have seen that this is problematic, since C is a physical object, whereas *m*(C) is a proposition. I shall argue below, from the premiss that modal realism is false, that there can be no perfect match between C and *m*(C).

In presenting this argument it is convenient to think about the whole world, so that we are concerned with the relation between the real world W and the maximal actual state of affairs (or 'actual world') *m*(W). We have already seen that W and *m*(W) cannot be identified, since W is a concrete object whereas *m*(W) is a logical entity. In spite of this, one might still think that the difference between W and *m*(W) is rather trivial, so that they can be regarded as exact replicas of each other. If this were so, however, then W and *m*(W) would be similar in ontological status. It would then follow that the other, non-actual possible worlds like *m*(W′), whose status is equal to that of *m*(W), would be just as real as W. This amounts to modal realism, which I take to be false.

The alternative to modal realism is to say that the possible worlds (including the one that is actual) are abstract, or ersatz. I do not want to discuss the exact meaning of 'abstract,' but I take it that the ontological status of possible worlds is somehow drastically inferior to that of the real world. Should we grieve over possible catastrophes that might have occurred but did not, in fact, happen? Is there anything that it is like to be a possible person? Do possible people suffer? I assume that the answer to each of these questions is No. Abstract objects, such as *m*(W), cannot be in perfect correspondence with real, concrete objects.

We still have not provided an answer to the question of how W and *m*(W) are related. This is a difficult task, but the following picture may help. Leibniz imagined a God who surveyed the possible worlds and, finding the best, chose it as the real world. Creating the real world cannot be just a matter of taking one of the possible worlds 'off the shelf' and renaming it as the real world, however. In that case, the possible worlds would have the same fundamental nature as the real world, which amounts to modal realism. Rather, the actual world, once selected, would have to undergo some mysterious process of 'concretization,' to turn it into the real world. The actual world *m*(W), on this view, is something like the *formal cause* of the real world W (in Aristotle's sense). The actual world helps to bring about the existence of the real world, but the two cannot be identified. The actual world is merely the pattern, structure, or skeleton of the real world.

To think that the actual world is identical to the real world is like thinking that the blueprint of a real aircraft is identical to that aircraft. Or, more precisely, it is like thinking that an aircraft blueprint can be turned into a real aircraft simply by writing 'really exists' somewhere at the bottom, or checking the box marked 'existent?' Existence is not like that.

### *3.3.7 Causal Interaction and Locality*

In view of the fact that C has no properties that are missing from *m*(C), the claim that *m*(C) is incomplete may seem rather intangible. Does this supposed incompleteness ever make a practical difference? I shall argue in chapters 6 and 7 that, in fact, the incompleteness of states of affairs is crucial in quantum theory – it is needed to make sense of experimental results. In this section I shall give a quick summary of the position argued for in those chapters.

As stated above, in a stochastic causal process the earlier states of affairs do not determine the later ones. The later states of affairs, and indeed the actual history of the process, emerge during the causal evolution. The causal process, in other words, not only actualizes possible states of affairs, but also 'chooses' which ones to actualize. It seems plausible to say that the states of affairs chosen in this way by a closed causal process **Z** concern only the process **Z**. A causal process on Mars, for example, cannot choose a state of affairs concerning Jupiter. This is not to say that events on Mars cannot cause events on Jupiter, for there can be causal processes such as light rays running from Mars to Jupiter. The point is that a causal process within a particular system can only actualize states of affairs pertaining to that system.

Suppose that two stochastic systems, **X** and **Y**, are each causally closed, so that every stage of **X** is a total cause of its successors, and the same for **Y**. In this case, as there are two causal processes, there must be two actual histories, one for **X** and one for **Y**. If, on the other hand, **X** and **Y** are in continuous causal interaction, then they form a single process ⟨**X**, **Y**⟩. Neither system has its own dynamical evolution; their causal developments are inextricably entwined. In such cases, the actual history produced is for the composite system ⟨**X**, **Y**⟩, which we shall call *m*⟨**X**, **Y**⟩. The crucial question then presents itself: Does the joint history *m*⟨**X**, **Y**⟩ split up into two separate histories, one for each subsystem?

A simple example of an interaction between two systems is where two particles interact by means of electromagnetic forces. The forces depend on the relative position and motion of the particles, so that the two form a single dynamical system. It is not possible to think about the dynamics of either particle in isolation from the other. We usually suppose, however, that each particle has its own trajectory, that is, its own sequence of positions in space. In classical (i.e., pre-quantum) physics, this assumption is always made. While interacting systems have to be considered as a whole, dynamically speaking, the history of the whole system consists of two separate histories, one for each subsystem.

In quantum physics, however, things are quite different. In chapter 6 I argue that an interacting pair of stochastic systems only has a joint history, in general; the individual systems may not have their own histories. My argument is based largely on the structure of quantum mechanics, and some experimental results, and will not be presented here. For now, let us just see how this might be true, and examine what it tells us about causation.

First, the principle stated above, that a closed stochastic causal process **Z** actualizes states of affairs concerning **Z** only, allows that the system ⟨**X**, **Y**⟩ may only have a joint actual history *m*⟨**X**, **Y**⟩. Since **X**, for example, is not a closed process, it may not have its own history. I see no good reason to suppose that, in general, mere parts of a causal process have their own histories. Note that I have no conceptual argument for the conclusion that parts of a causal process often do not have their own histories. My aim here is merely to show that this conclusion is possible.

Now suppose that, at time *t*, the systems **X** and **Y** break apart, and permanently cease their interaction. Let the state of affairs concerning ⟨**X**, **Y**⟩ at time *t* be $m_t$⟨**X**, **Y**⟩. Once they are physically separate, do the systems automatically acquire separate states of affairs? Again, I see no conceptual reason to suppose that they do. The separation of ⟨**X**, **Y**⟩ into **X** and **Y**, when the interaction ceases, is a *physical* matter, concerning the concrete systems. The separation of $m_t$⟨**X**, **Y**⟩ into $m_t$(**X**) and $m_t$(**Y**), by contrast, is a *logical* matter, concerning states of affairs. If $m_t$⟨**X**, **Y**⟩ does split up into $m_t$(**X**) and $m_t$(**Y**), then the relation between the three states of affairs is a logical one, namely, $m_t$⟨**X**, **Y**⟩ ⇔ [$m_t$(**X**) & $m_t$(**Y**)]. The joint state of affairs is the logical *conjunction* of the individual states of affairs. The relation between **X**, **Y**, and ⟨**X**, **Y**⟩ is (by contrast) not a logical relation. The system ⟨**X**, **Y**⟩ is the mereological sum of **X** with **Y**. It seems possible, therefore, that the systems **X** and **Y** might continue to lack their own histories, even after they have physically separated.

If $m_t$⟨**X**, **Y**⟩ is not the conjunction of $m_t$(**X**) and $m_t$(**Y**), then the joint state $m_t$⟨**X**, **Y**⟩ is obviously incomplete. After all, if $m_t$⟨**X**, **Y**⟩ were complete, then it would be just like ⟨**X**, **Y**⟩, and so would consist of two separate parts, as ⟨**X**, **Y**⟩ does at time *t*.

Let us suppose that $m_t$⟨**X**, **Y**⟩ is not equivalent to any conjunction of states of affairs, one concerning **X** only and the other concerning **Y** only. We can still use the symbols '$m_t$(**X**)' and '$m_t$(**Y**),' with their usual meanings, as $m_t$(**X**), for example, is the maximal actual state of affairs concerning **X** only. The conjunction $[m_t(\mathbf{X})\ \&\ m_t(\mathbf{Y})]$ exists, therefore, but contains less information than $m_t\langle\mathbf{X}, \mathbf{Y}\rangle$. The conjunction is not only incomplete, but also sub-maximal; it lacks available information. Now, if $m_t(\mathbf{X})$ is maximal concerning **X**, and $m_t(\mathbf{Y})$ is maximal concerning **Y**, then what is the information missing in $[m_t(\mathbf{X})\ \&\ m_t(\mathbf{Y})]$ about? It cannot be information about **X**, or information about **Y**.

The information that is missing is the relation between **X** and **Y**. This relation is actually *irreducible*, in the sense defined below.

**Definition** A relation between **X** and **Y** is state of affairs that concerns the aggregate $\langle\mathbf{X}, \mathbf{Y}\rangle$.

For example, the state of affairs of Jack being taller than Jill is a relation between Jack and Jill. Many relations are reducible, in the sense that they are entailed by conjunctions of states of affairs, where each conjunct concerns one of the systems only. The relation of Jack being taller than Jill is entailed by (Jack is 6′ tall & Jill is 5′8″ tall), for example. An irreducible relation, on the other hand, is defined below.

**Definition** A relation between **X** and **Y** is *irreducible* iff it is not entailed by $[m(\mathbf{X})\ \&\ m(\mathbf{Y})]$.

Thus, if $m_t\langle\mathbf{X}, \mathbf{Y}\rangle$ is not the conjunction of $m_t(\mathbf{X})$ and $m_t(\mathbf{Y})$, then there is an irreducible relation between **X** and **Y**. We see that the incompleteness of the maximal state of affairs $m_t\langle\mathbf{X}, \mathbf{Y}\rangle$ entails that there is an irreducible relation between **X** and **Y**.

There are other ways to describe this situation. Since conjunction is sometimes described as the logical *product*, we might say that the joint state of affairs $m_t\langle\mathbf{X}, \mathbf{Y}\rangle$ does not 'factorize' into the conjunction $[m_t(\mathbf{X})\ \&\ m_t(\mathbf{Y})]$. Also, since the state of affairs $m_t\langle\mathbf{X}, \mathbf{Y}\rangle$ concerning the whole system $\langle\mathbf{X}, \mathbf{Y}\rangle$ is not entailed by the maximal states of affairs concerning its parts, we might say that the state of affairs $m_t\langle\mathbf{X}, \mathbf{Y}\rangle$ is logically *emergent*.

### *3.3.8 Determinacy*

The distinction between determinism and determinacy is familiar. To say that an event is *determinate* simply means that it is well defined, definite, sharp, and so on. This falls far short of saying that the event's occurrence is predictable (in principle) from earlier events. The notion of indeterminacy is essential to understanding the ideas in this chapter, so we shall conclude with a few remarks about it.

It is hard to say exactly what is meant by an event's being determinate, as no alternative seems conceivable. How could reality *itself* be indeterminate, or 'fuzzy'? My knowledge of an event may be fuzzy, to the extent that the nature of the event is not fully known by me, but how could the event itself be indeterminate? The idea developed in the previous two sections, however, that even maximal states of affairs are incomplete, provides us with an understanding of indeterminacy. To say that an event C is indeterminate is to say that its maximal state of affairs, *m*(C), is incomplete. An indeterminate event has no exact logical counterpart.

In this sense, all physical events are indeterminate, so the term may seem to have little value. In view of the symmetry-conservation argument, however, it is important to recognize that indeterminism depends on indeterminacy in this sense. If every event in a system were fully determinate, having a complete state of affairs, then the system would be deterministic. Also, given that some systems are more deterministic than others, it is likely that indeterminacy is a matter of degree, so that some events are more determinate than others. In particular, we shall see in chapters 6 and 7 that some systems can even be indeterminate with respect to standard physical properties, such as position and momentum.

Objective indeterminacy is also needed for there to be irreducible relations, logically emergent properties, and so on. It is no accident that quantum mechanics, which presents the world as being genuinely stochastic, also seems to involve irreducible relations and objective indeterminacy.

# 4

# Physical Chance

In this chapter I shall use the ideas developed in chapters 2 and 3 to state the causal theory of chance. There is nothing new about the basic idea, which is similar to a previous proposal,[1] but there are some innovations in matters of detail.

## 4.1 THE DEFINITION OF CHANCE

Like Lewis (1980: 109), I think that Miller's Principle[2] shows us how to proceed in giving a theory of chance. Miller's Principle, in its simplest form, states that the epistemic probability of some event $A$, given only that its physical chance is $p$, is also $p$. Thus we have

***Miller's Principle***

If one knows that the chance of $A$ is $p$, then this authorizes a degree of belief $p$ in $A$.

I write 'in its simplest form,' as the principle thus stated has counter-examples. Such cases arise when one's epistemic state $K$ has too much extra knowledge about $A$, in addition to knowing that $P(A) = p$. For instance, if $K$ includes knowledge that $A$ is true, then the principle fails. This complication is discussed in §4.3 below.

The principle is important as it provides a link between the two kinds of probability, physical and epistemic. It shows us that the two kinds cannot be fundamentally distinct; one must be a special case of the other, or both are linked to some third, more basic, kind. The situation is analogous to the fact that, in Newtonian mechanics, the gravitational mass of a body is exactly proportional to its inertial mass. For Newton this was a brute fact that has no explanation, as the concepts of inertial and gravitational mass are totally different. The fact that they are proportional, however, shows that in fact they must be two manifestations of the same property, and so it turns out. In general relativity, only gravitational mass is required, and we can see how it determines a body's resistance to acceleration relative to the 'fixed stars.'

If chance and epistemic probability are connected, then what is the link? Either one of them can be reduced to the other, or they are linked via some third thing. Epistemic probability cannot be a special case of chance, for it depends on the knowledge one possesses. However, it seems equally clear that chance cannot be a special case of epistemic probability, for the latter is very much bound up with human brain structure. The first constraint on physical chance, laid down in chapter 1, was that chances depend only on the physical properties of the system in question.

The third alternative is that epistemic and physical probability are linked via their being related to some third thing; this is the solution I propose. The third thing is logical probability. It is reasonable for chance to be a special case of logical probability, since the latter is independent of all contingent matters of fact. Thus, if the information given to the logical probability function is entirely concerned with physical properties of the system in question, then the logical probability of any event in this system, given those facts, is determined by those properties alone. Moreover, there is a link between logical and epistemic probability, namely, the Authority of Logic principle.

We have seen that the logical-probability function is a function of two arguments, whereas chance is a function of one argument. So, if we are to reduce chance to logical probability, we need to find something to fill the second argument place. Our task is to discover the state of affairs, $U$ say, such that $P(A) = Pr(A \mid U)$. It would, as stated above, be quite intolerable if the chance function depended on human beings in some way. It could not then be described as 'physical,' or 'objective.' Thus, the choice of $U$ cannot depend on what any human knows about the system. On the contrary, $U$ must be fixed by only physical facts about the system.

As a first guess, we might say that $U$ contains complete knowledge of the system in question, as chance should not depend upon ignorance in the way epistemic probability does. This clearly will not do, however, as $A$ is also a state of affairs concerning the system, and so either $A$ or $\neg A$ will be a consequence of $U$, making all chances either 0 or 1. Clearly, we cannot tell the logical probability function everything, so what information do we give it?

As stated in chapter 1, the answer is to supply the logical probability function with a maximal specification of the *efficient causes* of the actual history of the system, of which there are two. First there is the dynamical nature of the system, represented by the generalized lagrangian[3] $\mathscr{L}_{\mathbf{X}}$, and second there is the boundary condition $bc_{\mathbf{X}}$. The boundary condition is a constraint on the motion of the system that is independent of the system's dynamical nature.

It may not be clear that the boundary condition for a system is a cause[4] of its motion, or actual history. Indeed, although the boundary condition is a constraint of some kind, it is sometimes considered to be an *epistemic* constraint, rather than a physical one. In other words, the boundary condition may be viewed as a piece of *information* about the system that constrains what we may believe about the motion, but does not physically constrain the motion itself. This attitude toward the boundary condition is strongly encouraged by the fact that, for a deterministic system, the possible boundary conditions are grouped into equivalence classes, where the members of an equivalence class all generate the same actual history. Given this fact, how can it be maintained that just one of the boundary conditions is the real one, the cause of the motion, whereas the others in that equivalence class are not?

The full answer to this question must wait until §5.6, where the matter is discussed in detail. The basic idea is that the grouping of possible boundary conditions into equivalence classes is a special feature of deterministic systems that does not occur for any stochastic system. In a stochastic system, distinct boundary conditions give rise to different chance functions, and hence (probably) to different motions. In that section, certain physical phenomena are explained by a hypothesis about the nature of the boundary condition. In other words, the boundary condition is postulated as a cause of the motion, and so it is necessary to regard the boundary condition as a physical constraint.

The causal theory of chance is then as follows:

**4.1.1 Definition** (The causal theory of chance)

> If $\mathscr{L}_{\mathbf{X}}$ represents the dynamical nature of a system **X**, and $bc_{\mathbf{X}}$ its boundary condition, then $P_{\mathbf{X}}(A) = Pr(A \mid \mathscr{L}_{\mathbf{X}} \;\&\; bc_{\mathbf{X}})$.

Note that '*A*' here denotes a possible state of affairs, rather than a concrete event. In the general, stochastic case, as well as the deterministic case, $\mathscr{L}_{\mathbf{X}}$ does not contain any modal facts, but is just an ordinary physical state of affairs. As we saw in the previous chapter, the logical probability function does not need to be given probabilities in order to yield probabilities. In other words, we may say that probability enters the physical world not by there being probabilistic physical facts, but by there being relations of partial entailment between ordinary physical facts. If this were not so, and the state of affairs $\mathcal{L}_{\mathbf{X}}$ made probabilistic assertions, then we would not have any theory of chance at all, as the analysis would be circular.

The chance function thus defined will satisfy the axioms of probability, as it is just the logical probability function supplied with particular information.

## 4.2 CHANCE IS RELATIVIZED TO A SYSTEM

It should be noted that, in Definition 4.1.1, the chance of an event *A* is defined for a particular system **X** in which *A* occurs. On this view, therefore, there is no such thing as the chance of *A simpliciter*; one must always specify a reference system, one that includes the event *A*.[5] If we assume (as I do in the next chapter) that the composition of two systems is also a system, then every event occurs in many different systems. If an event *A* occurs within **X**, and **X** is a subsystem of **Z**, then *A* also occurs within **Z**.

As an illustration of the fact that chances are system-dependent, consider a system **X** that consists of a fair, six-sided die together with a device that rolls the die whenever a button is pushed. The button is pushed on some occasion, let us suppose, causing some outcome *A*, such as a '4' on the die. Since the die is fair, it seems clear that the chance of this outcome within **X** is $\frac{1}{6}$. Now let us consider a larger system **Z**, however, which includes **X** as a subsystem. The button in **X** was pushed by some part of **Z**, another chancy contraption that lies outside **X**. If the chance (within **Z**) of the button being pushed at all is only $\frac{1}{2}$, then it appears that the chance of *A* within **Z** is only $\frac{1}{12}$, not $\frac{1}{6}$. We thus write $P_{\mathbf{X}}(A) = \frac{1}{6}$, and $P_{\mathbf{Z}}(A) = \frac{1}{12}$, to mark this difference.

The fact that chances are defined only for a particular reference system is related to the idea that the chance of a single event can change with time. Consider, for instance, a particle undergoing a random walk within a maze, and let the event *A* be that the particle emerges from the exit of the maze before some given time, say noon. It may be that at eleven o'clock, perhaps, the particle is only a few steps away from the exit, and so at this time the chance of *A* is quite high. Unfortunately, by 11:30 am the particle has moved back toward the centre of the maze, so that its chance of escape before noon is now much diminished. As is shown in §5.6, this time-dependence of chance can be accounted for under the assumption that a time slice[6] of a system is itself a system, possessing its own lagrangian and boundary condition. The chance of an event *A* at time *t* within **X** can then be defined as the chance of *A* within some time slice $[t, \tau]$ of **X**, where $\tau$ is any time after *A* has finished.

## 4.3 LEWIS'S OBJECTIONS

David Lewis (1980) does not propose a definite analysis of chance, but presents a range of possible approaches that includes my own. By a rather circuitous route, which I shall not retrace here, Lewis arrives (1980: 97) at the following statement, which he considers to be a form of Miller's Principle:

$$P_{tw}(A) = P_K(A \mid H_{tw} \,\&\, T_w).$$

The subscripts will require some explanation. The chance function is written $P_{tw}$ rather than $P$, as Lewis considers chance to be both time-dependent and world-dependent. These complications do not concern us here. The proposition $H_{tw}$ is a complete description of the history of world $w$ up until time $t$, and $T_w$ is a complete 'theory of chance' for the world $w$. A theory of chance is a conjunction of conditionals from history to chance, that is, a specification of what the chances of events would be for each possible initial segment of $w$'s history.

Lewis then considers the suggestion that this formulation of Miller's Principle could serve as an analysis of chance. Clearly, such an analysis would bear strong similarities to the causal theory proposed above in Definition 4.1.1, so let us look at the difficulties Lewis raises against this idea. He sees two separate problems. First, the appeal to $P_K$, which Lewis calls a 'reasonable initial credence function,' will be illegitimate unless we can give an account of it that does not itself appeal to chance. Lewis believes that such an account will probably require symmetry constraints in order to be sufficiently restrictive about what counts as 'rational,' but these constraints are associated with well-known problems. He points out, for instance, that 'it is not possible to obey too many different restricted principles of indifference at once and it is hard to give good reasons to prefer some over their competitors' (1980: 111). I have argued in chapter 2, however, that there is a symmetry principle that seems to be a part of logic, and free of paradox, so I shall not discuss this objection again here.

The second problem is that such an analysis runs the risk of circularity, on account of the use of $T_w$. For this term, which appears in the analysans, itself involves the concept of chance. In order to avoid this objection, Lewis notes (1980: 111–12), we must replace $T_w$ with a 'Humean' statement – one that only describes matters of particular fact, and not anything modal. The only possible way to accomplish this, Lewis believes (correctly I think), is if each world has the same theory of chance. This would make the theory of chance necessary rather than contingent.[7]

According to the causal theory of chance, the history-to-chance conditionals are provided by the dynamical properties of the system, together with logical probability. The dynamical properties $\mathscr{L}_{\mathbf{X}}$ are independent of the actual history of **X** – in Lewis's terms they are the same in every nomically possible world. Moreover, the same relations of logical inference hold in all worlds. Thus, my causal account of chance is one according to which the 'theory of chance' is necessary.[8] In his discussion, Lewis does not refer to intrinsic dynamical properties of systems, such as geometrical shape, viscosity, elastic constants, masses, charges, and so forth, so it is not clear whether he is willing to countenance such things.[9] Their importance to my theory is highlighted by Lewis when he points out that

> according to [the idea that the theory of chance is necessary] if I were perfectly reasonable and knew all about the course of history up to now (no matter what that course of history actually is, and no matter what time is now) then there would be only one credence function I could have. Any other would be unreasonable. It is not very easy to believe that the requirements of reason leave so little leeway as that. (1980: 113)

Indeed, it is not very easy! This is surely a huge understatement, as (from my point of view) it makes the dynamical properties of a system logically necessary! I am not sure what it would even mean to assert that this pendulum has a string-length of 28.5 cm by logical necessity. If one grants the existence of $\mathscr{L}_{\mathbf{X}}$, however, then one's credence is not stretched nearly so far.

## 4.4 A PROOF OF MILLER'S PRINCIPLE

Since the causal theory of chance was motivated by the desire to account for Miller's principle, to make it not a mere coincidence, we should ensure that the principle holds for it. Consider some event *A*, about which the only relevant information you have is that the chance of *A* is *p*. Given only this knowledge, is the epistemic probability of *A* also *p*? I shall now show that it is, using two premisses. I need first a screening-off assumption, and second the Authority of Logic principle of §2.2.3.

If we let *U* be the state of affairs $\mathscr{L}_{\mathbf{X}}$ & $bc_{\mathbf{X}}$, where **X** is any system in which *A* occurs, then the screening-off assumption is that $K_U$ is an authority for one's own epistemic state. In other words (since $K_U$ is obviously veridical), we assume that $K_U$ is superior to one's own state. This assumption will not always be true, of course, but then (as noted in §4.1) Miller's Principle has counter-examples as well. The derivation of Miller's Principle will be adequate, therefore, if these two sets of counter-examples coincide exactly. That is, we require that Miller's Principle hold in all and only those cases where the knowledge in $K$ is screened off by knowledge of $\mathcal{L}_{\mathbf{X}}$ & $bc_{\mathbf{X}}$.

**4.4.1 Theorem** (Miller's Principle)

Let $U = \mathcal{L}_{\mathbf{X}}$ & $bc_{\mathbf{X}}$. Then, if $K_U \geq K$, then $P_K(A \mid P_{\mathbf{X}}(A) = p) = p$.

*Proof*: By the causal theory of chance, $P_X(A) = p$ means that $Pr(A \mid U) = p$. Then, supposing that $K_U \geq K$, we have that $K_U$ is an authority for $K$, so that $P_K(A \mid Pr(A \mid U) = p) = p$. Thus, $P_K(A \mid P_X(A) = p) = p$, as required.■

We see that, since chance is defined using logical probability, it is also authoritative in the same way. One might therefore call Miller's Principle the *Authority of Chance* principle. It should be noted that this conditional form of the principle answers the previously difficult question[10] of which epistemic states Miller's Principle applies to. Any proposition is admissible, in Lewis's sense, provided it is screened off by $\mathcal{L}_{\mathbf{X}}$&$bc_{\mathbf{X}}$.

## 4.5 THE OBJECTIONS OF HOWSON AND URBACH

Howson and Urbach (1993: 342–4) criticize a theory of chance that they simply call 'the theory of objective chance.' According to this theory, a stochastic system (such as a coin in a coin-tossing experiment) has a certain chance of producing each of its possible outcomes. Thus, for instance, there is a certain chance that the coin will land heads, and this number attaches to a single trial, rather than to a collective of outcomes. The problem of showing that the chancc function is a measure is solved by postulating that Miller's Principle holds, so that 'chances license numerically equivalent degrees of belief.' By tying chances to coherent degrees of belief, which (according to a plethora of arguments) must obey the axioms of probability, it follows that chance must be a probability function as well.

Howson and Urbach are not critical of the theory as described so far, but rather of Levi's additional claim (Levi 1980: 258) that no characterization of chances independently of Miller's Principle is either possible or necessary. Chances are left then as mysterious, postulated entities that justify corresponding degrees of belief. As they point out, this is rather unsatisfying and we should try to do better. Howson and Urbach also attack Lewis on the same grounds, although less justly since Lewis does not actually offer any theory of chance, and hence does not offer this theory. Lewis does however place the following constraint on any proposed theory: 'I would only ask that no such analysis [of chance in terms of epistemic probability or relative frequency] be accepted unless it is compatible with the Principal Principle' (1980: 90). As a necessary condition for a theory to be acceptable this is beyond reproach, but as a sufficient condition ('I would only ask ...') it is much too weak, and does leave him open to a related objection. For if an analysis of chance is merely compatible with Miller's Principle, and does not entail it, then the latter is left unexplained. As I argued in §4.1, this would be like the unexplained numerical equality of inertial with gravitational mass in Newtonian mechanics. An adequate theory of chance should explain why, in certain cases, chances and epistemic probabilities are numerically equivalent.

The causal theory of chance faces no objection along these lines since, as is shown in §4.4, Miller's Principle may be derived from Definition 4.1.1

## 4.6 CHANCE AND RELATIVE FREQUENCY

If there is such a quantity as physical chance, then it is surely connected in some way with relative frequency. More precisely, for a given experiment that has $A$ as one of its possible outcomes, there is surely some link between $P(A)$ and the relative frequency of outcomes similar to $A$ in a large class of similar experiments. In this section it will be shown that, according to the causal theory of chance, there are two links between chance and relative frequency.

The first link between chance and frequency is epistemological. The chance of an event may be inferred, albeit approximately and fallibly, by measuring the relative frequency of similar events in a large class of repeated trials. Chance, in other words, is *empirically accessible*, in that it can be discovered (approximately and fallibly) by empirical means.

The second link between chance and frequency is concerned with explanation. Chances are often cited as explanations of a certain physical phenomenon, namely, the tendency of an outcome type of a given kind of trial to be associated with a stable relative frequency. In quantum mechanics, for instance, one explains the distribution of actual positions of photons detected at a screen by positing a chance distribution for each single photon. One says that the density of detected photons is greatest at this point *because* that is where the chance of detection is highest for each photon.

To show that the causal theory of chance accounts for both of these connections between chance and frequency, we need first to define some terms. The standard experimental situation in which frequencies are discussed is when a certain type of experiment is performed many times. We can either suppose that the same apparatus is used for each experiment, being completely reset before each new trial, or that many exactly similar sets of apparatus exist. In the former case the experiments are well-ordered by the times at which they occur, and in the latter case we shall suppose that some other well-ordering exists. Let $K$ be the set of possible outcome-types on each experiment, and $F$ some field of subsets of $K$. Then the combined outcome of all the trials together may be represented by an outcome sequence such as $\mathbf{s} = \langle s_1, s_2, ..., s_n, ...\rangle$, where each $s_i$ is a member of $K$. In the case where the total number of experiments is some finite number $n$, we can define the relative frequency of each member $A$ of $F$ as the number of times $A$ occurs divided by $n$. Writing the relative frequency of $A$ in $\mathbf{s}$ as $f(A, \mathbf{s})$, we have that

$$f(A, \mathbf{s}) = (\#A \text{ in } \mathbf{s})/n.$$

If the number of experiments is countably infinite, then we can still define $f_n(A, \mathbf{s})$ as the relative frequency of $A$ in the first $n$ terms of $\mathbf{s}$, and then we define

$$f(A, \mathbf{s}) = \lim_{n\to\infty} f_n(A, \mathbf{s}).$$

This limit only exists for some outcome sequences, of course. In cases where the sequence $f_n(A, \mathbf{s})$ does not converge, the quantity $f(A, \mathbf{s})$ is not defined. In circumstances where it is clear which outcome sequence is in question we shall write '$f(A)$' instead of '$f(A, \mathbf{s})$.'

Now let us look at the issue of whether chance is empirically accessible. Suppose an experiment is performed and a certain outcome obtained, say $A$. If one desires to know what the chance of that outcome was, one can repeat the entire experiment *ab ovo* a very large number of times and record the outcome in each case. If one goes to the trouble of calculating the relative frequency of As in the first $n$ trials, for every $n$, then it usually appears that the values are converging towards a limit.[11] The experimenter then has a high degree of confidence that the chance of $A$ lies in a certain interval centred on the relative frequency of $A$ for the entire class. The problem is to show that this reasoning is justified, if the causal theory of chance is correct.

The reader may have noticed that the concept of relative frequency does not appear in Definition 4.1.1, so one may doubt that the theory can account for the fact that chances are empirically accessible. In this section I shall show that such fears would be unfounded.

The first thing to realize is that, according to generally accepted ideas about chance, a statement about the chance of an event is not logically equivalent to any statement that is purely about frequency.[12] This is quite clear for frequencies in finite sequences of outcomes, but is also true for countably infinite sequences of trials. Consider, for instance, a fair coin that is flipped ten times. The chance of heads is $\frac{1}{2}$ on each toss, but the relative frequency of heads may or may not be $\frac{1}{2}$. The relative frequency of heads may take any value in the list $0, \frac{1}{10}, \frac{2}{10}, \ldots, \frac{9}{10}, 1$. If we assume that the individual trials are pairwise probabilistically independent,[13] then using the axioms of probability we can calculate the chance of each possible relative frequency in this set of experiments. It turns out that the most likely frequency is $\frac{1}{2}$, with a chance of about 0.25.

Now suppose we increase the number of trials from ten to one thousand. Again, the relative frequency of heads could be anything from 0 to 1 in steps of $\frac{1}{1000}$, but now there is a 95 per cent chance that the frequency will lie between 0.47 and 0.53. The general picture is that as the sequence of trials is lengthened, the chance distribution over possible frequencies tends to a Gaussian normal distribution, which becomes increasingly peaked around the physical chance. What happens then for a countably infinite sequence of experiments?

In the infinite case, we are of course concerned with the limiting relative frequency. One result, known as the Strong Law of Large Numbers, states that (for our coin-tossing experiment) the limiting relative frequency almost surely exists and is equal to $\frac{1}{2}$. If one enquires about the meaning of 'almost surely,' then the usual answer is that an event occurs almost surely if and only if its chance is one. It is important to realize that events with probability one need not occur, that is, they are not quite certain, or necessary – that is why such events are described as *almost* sure. In order for a proposition $A$ to get a probability of one, the contract [$1 if $A$] need not be worth $1. The contract need only generate a Dedekind cut on the scale of rational values, whose upper set has an infimum of 1 (see Def. 2.3.4).

The Strong Law of Large Numbers does not *entail*, therefore, that the limiting relative frequency will equal the chance. It is possible, for example, that the outcome sequence will consist entirely of tails, so that the relative frequency of heads is zero. Thus, since the statement that $P(H) = \frac{1}{2}$ does not (together with the independence assumption) entail that $f(H) = \frac{1}{2}$, it is impossible to regard the two statements as equivalent. As we shall see in section 4.7.1, this fact is the Achilles heel of all frequency theories of objective probability.

The Strong Law of Large Numbers is an example of what I call a chance-chance statement. It is a conditional, where both the antecedent and the consequent are about chances. It may be expressed in the form 'if $P(A) = p$, then $P(f(A) = p) = 1$.' The conclusion so far about the link between chance and frequency is that there are no true chance-frequency statements. That is, there is no true statement of the form 'If $P(A) = p$, then necessarily $f(A) = p$,' or 'If $P(A) = p$, then if an infinite sequence of trials were performed, $f(A)$ would (necessarily) equal $p$,' and so on.

(1) chances are empirically accessible, by measuring relative frequencies, and (2) hypotheses about chance may be used to explain observed relative frequencies.

## 4.7 FREQUENCY THEORIES OF PROBABILITY

There are other interpretations of the probability calculus in terms of purely physical properties, which may be described as frequency theories of probability. The first sophisticated formulation of this idea is due to Richard von Mises, but many others have followed in his footsteps, producing their own versions. Among these we shall consider those of Popper, Howson and Urbach, and van Fraassen.

Frequency theories are founded upon two common observations. The first is that relative frequencies are, formally speaking, probabilities.[20] In any outcome sequence **s**, for instance, if $A$ and $B$ are mutually exclusive, then $f(A \vee B, \mathbf{s}) = f(A, \mathbf{s}) + f(B, \mathbf{s})$, which is an axiom of probability. The other axioms also hold, including even the Principle of Conditioning, as we can define $f(A \mid B)$ as the relative frequency of $A$ in the sub-sequence of terms that are members of $B$. Second, there is a close link between physical probability and frequency, which we examined in the previous section.

There are two basic kinds of frequency theory that, following van Fraassen (1980: 181, 190), I shall call *strict* frequency theories and *modal* frequency theories. A strict frequency theory, such as that proposed by Reichenbach (1935), states that the objective probability of an event-type is its relative frequency in the actual sequence of outcomes $\mathbf{s}_a$, that is, $P(A)$ is just $f(A, \mathbf{s}_a)$ by definition. The proper response to this view seems to be as follows. It cannot be denied that relative frequencies among outcomes of actual experiments are objective quantities, and moreover that they are probabilities in the formal sense. Thus, one might well call them 'objective probabilities.' They are not, however, what we mean by physical chances, as was shown in §4.6. It is an essential property of chance that it does not strictly determine a unique relative frequency, but merely provides a chance distribution over the various possible relative frequencies. In short, frequency is one thing and chance is another.

Reichenbach cannot claim that he provides an analysis of our common notion of chance, although he may hold that that notion is bankrupt, and that his alternative is sufficient for the needs of science. In response here I would say first that the notion of chance is not bankrupt, since the causal theory gives a good account of it. Second, we cannot do without chances, as they are so deeply involved in quantum mechanics.

The second kind of frequency theory, the modal frequency theory, is far more plausible and has many distinguished advocates. There are many different versions of it, however, which we shall consider in turn.

### *4.7.1 Von Mises*

Von Mises's frequency theory (1928) starts with the idea of a collective. A collective is an infinite outcome sequence **s** with the following two properties.

1 The Axiom of Convergence: $f(A, \mathbf{s})$ exists for each $A$ in $F$.
2 The Axiom of Randomness: If we take any sub-sequence $\mathbf{s}'$ of **s**, which is picked out by a $1-1$ increasing recursive function, then $f(A, \mathbf{s}')$ also exists and is equal to $f(A, \mathbf{s})$.[21]

Now let us consider some experimental set-up, and one possible outcome $A$ for that experiment. The objective probability of $A$, which we shall write $P(A)$, exists if and only if

1 this set-up would (on an infinite sequence of experiments) produce some collective or other, and
2 any two collectives $\mathbf{s}_1$ and $\mathbf{s}_2$ that might be produced are such that $f(A, \mathbf{s}_1) = f(A, \mathbf{s}_2)$, for every $A$ in $F$.

If these two conditions are satisfied, then we can define $P(A) = f(A, \mathbf{s})$, where **s** is any collective that might be produced by the experiments. (I assume that von Mises's claim is that an experiment for which objective probabilities are defined necessarily produces some collective or other, but not necessarily any collective in particular.)

It is clear that this definition is in conflict with the common-sense notion of chance, as according to von Mises it is impossible for a fair coin, that is, one with $P(H) = \frac{1}{2}$, to land heads every time in an infinite set of tosses. The term '$P(H)$' is not even defined unless the coin necessarily generates a collective. Moreover, it is hard to see how the collective $\langle H, H, H, ...\rangle$ could be impossible if, on each individual trial, the outcome $H$ is possible for that trial. It would seem to require that some of the experiments, at least, 'know' the outcomes of other experiments, even though they are supposed to be carried out independently. Worse still, it is not just the collective $\langle H, H, H, ...\rangle$ that is excluded for a process with $P(H) = \frac{1}{2}$, but an uncountably infinite set of other collectives. And then, of course, the uncountably infinite set of non-collectives is also ruled out. On the face of it at least, and according to standard probability theory, each of these outcomes is individually just as possible as any one sequence in the select group of collectives. Let us call this problem, that frequencies do not *necessarily* equal probabilities, even in an infinite class of trials, the *cardinal problem* for frequency theories.

Excluding 'communication' between experimental trials, the only way to restrict the outcome sequence to a select group of collectives is by the outcomes, as a group, being determined by something or other. This idea is also extremely problematic, however, for two reasons. First, it would mean that the first toss is determined perhaps to land heads, the second heads also, the third tails, and so on, even though the experiments are supposed to be set up under maximally similar initial conditions. This is a contradiction, so we would have to say that the initial conditions vary slightly from experiment to experiment. Then, however, it is clear that these mysterious variations are faced with the same difficult task of guaranteeing that the outcome sequence is a collective with the right frequency. They cannot be 'random,' but must be rigged to follow some preset pattern. There must be something in the background, pulling the strings so to speak.

Second, if the trials are all deterministic, then one is inclined to agree with Lewis (1980: 117–21) that the genuine chance of heads varies from trial to trial, being either 0 or 1 in each case. The number $\frac{1}{2}$ is then some sort of 'counterfeit chance,' depending upon human ignorance. It is not determined by the experimental arrangement itself.

This objection (that chance set-ups would not necessarily produce collectives upon infinite repetition, let alone collectives with the right frequencies) seems to be quite fatal to all modal-frequency theories, so it is surprising that it is so rarely mentioned in the literature. The only reason I can see for ignoring it is as follows. In recent discussions of the measurement problem in quantum mechanics, there is much talk of two quantities being equal FAPP, that is, For All Practical Purposes.[22] For instance, if a density matrix is such that it is practically impossible to distinguish it experimentally from a classical mixture, then one says that the state is a classical mixture, FAPP. To many who work on the experimental end of the subject, this seems a completely satisfactory solution to the measurement problem. The more theoretically inclined, however, feel very uneasy about it.

The situation with chance seems rather similar. For any sensible person, the distinction between events that are necessary and those that occur almost surely is too fine to see. If two propositions are equivalent almost surely then they just are equivalent, FAPP. Anyone who insists on a difference here is just being perverse.

Am I being perverse? I think it is important to distinguish between practical and theoretical contexts. In a practical context one can normally treat an event whose probability is as low as 0.9999 as being certain to occur, and it would indeed be perverse to express doubt about an event that occurs almost surely. In a theoretical context the situation is quite different. It is a matter of simple logic. If two propositions are equivalent, so that one can be analysed as the other, then they entail one another. Therefore, since the statement '$P(H) = \frac{1}{2}$' does not entail 'if the experiment were repeated infinitely many times, we would necessarily have $f(H, \mathbf{s}) = \frac{1}{2}$,' they are not equivalent. In theoretical contexts, a miss is as good as a mile.

It should be noted that the cardinal problem is universally recognized with regard to defining objective probability as relative frequency in a *finite* sequence of outcomes. No one, for instance, defines $P(H)$ as the relative frequency of heads that would exist in a sequence of 1000 tosses, since there is no single relative frequency that would necessarily occur in this experiment. The use of an infinite experiment constitutes a 'FAPP' solution to the cardinal problem. The unwanted relative frequencies are ignored by making them very, very improbable.

A more common objection is that von Mises's probabilities cannot be determined empirically. Let us call this the problem of *empirical access*, which is also common to all frequency theories. The problem is that the limit of a sequence depends only upon the infinite final segments, so to speak, and is entirely independent of any finite initial segment. Yet, of course, only finite initial segments are open to examination. More precisely, the convergence of the outcome sequence is what is known as a *tail event*,[23] that is, the limit of the sequence $f_n(A, \mathbf{s})$ is unaffected by any change in the first $r$ terms of $\mathbf{s}$, for any finite $r$. Thus, for any finite $r$, the first $r$ terms of $\mathbf{s}$ tell us precisely nothing about the value of $f(A, \mathbf{s})$.[24]

Another way to express this objection is to consider what we might call *deceitful collectives*. A deceitful collective, with $f(A) = \frac{1}{2}$ say, is one that seems to be converging on some other relative frequency for (say) the first billion terms. For instance, it may be converging quite convincingly to $\frac{3}{4}$ in the first billion terms, but then suddenly switch and converge to $\frac{1}{2}$ thereafter. Since whether or not an outcome sequence is a collective depends only upon its limiting behaviour, such 'deceitful' sequences are perfectly good collectives. We may contrast deceitful collectives with *honest* ones, that is, ones that begin to converge to their final limit right from the start. Now suppose we repeat a type of experiment one billion times, and obtain a relative frequency that is close to $\frac{3}{4}$. Can we take this as very good evidence that $P(H)$ is close to $\frac{3}{4}$?

To keep things simple we shall compare two hypotheses, that $P(H) = \frac{3}{4}$ and that $P(H) = \frac{1}{2}$. To be fair, we shall assume that these have the same initial (epistemic) probability. The question is now whether this evidence of one billion trials causes the hypothesis that $P(H) = \frac{3}{4}$ to become much more probable than its alternative. The crucial terms in a Bayesian evaluation of this question are $P_K(f_n(H, \mathbf{s}_a) = \frac{3}{4} \mid P(H) = \frac{3}{4})$ and $P_K(f_n(H, \mathbf{s}_a) = \frac{3}{4} \mid P(H) = \frac{1}{2})$. For long-run frequencies to be empirically accessible, the first term must be much greater than the second.

Intuitively, the ratio of these two quantities depends upon the relative probabilities of two classes, the deceitful collectives, $\mathbf{s}'$, with $f(H, \mathbf{s}') = \frac{1}{2}$ (which converge to $\frac{3}{4}$ in the first billion terms) and the honest collectives, $\mathbf{s}$, with $f(H, \mathbf{s}) = \frac{3}{4}$. All of these sequences yield the observed datum. Now, since the two classes of collectives, those with $f(H) = \frac{1}{2}$ and those with $f(H) = \frac{3}{4}$, are assumed to have equal prior probability, the question is about the relative probabilities of honest and deceitful collectives within classes of fixed $f(H)$. Von Mises requires that, among collectives with $f(H) = \frac{1}{2}$, for instance, the class of honest sequences is much more probable than the class of deceitful ones.

This difference of probability cannot be a matter of cardinal power, for both classes have the power of the continuum. Moreover, since von Mises has nothing in his account to suggest any privileged measure over the set of collectives, it is hard to see where it could come from. This problem of empirical access therefore appears to be quite fatal to von Mises's account.[25]

### *4.7.2 Popper*

Popper's theory of objective probability (Popper 1934) is almost the same as that of von Mises. The only difference seems to be that Popper is willing to ascribe probabilities to particular events, rather than merely to types of event. Popper, like von Mises, defines the probability of an event-type as the relative frequency of that event-type in all the collectives that might be generated by an infinite repetition of the experiment. Now, for Popper, these imagined repetitions are supposed to be carried out under exactly similar conditions; thus, assuming $f(A, \mathbf{s})$ differs from 0 and 1, it follows that the outcomes are irreducibly indeterministic. (If they were deterministic, then the same initial conditions would invariably lead to the same outcome.) Thus, the probability is determined entirely by the experimental conditions on a single trial: these conditions determine the possible collectives that might be generated upon infinite repetition, and the collectives determine the probability.

For this reason, Popper describes his theory as one according to which there are objective, single-case probabilities. Such a probability is seen as the propensity for experiments of that type to produce collectives having that particular relative frequency. Popper seems to be quite correct about this. The puzzle is why von Mises and others insist that probability is irreducibly a property of the whole collective, and cannot be attached to a single event.

Howson and Urbach (1993: 340) do give an argument, on von Mises's behalf, as to why probabilities are not defined for a single experiment. They point out that, in a coin-tossing experiment for example, there are many causes that combine to produce the outcome in a single case. In addition to the intrinsic properties of the coin there are variations in air density, convection currents, the precise way in which the coin is released, and so on. This is quite true, of course. They then assume, however, that the outcome of a toss is uniquely determined by a set of parameters $q_1$, $q_2$, ..., $q_k$. This leads to a serious problem for Popper, as it means that an exact repetition of an experiment will yield the same result. It then follows that a single experiment can only define one of the two collectives $\langle T, T, T, ...\rangle$ and $\langle H, H, H, ...\rangle$, and so all single-case probabilities are either 0 or 1.

The way to avoid this unwelcome triviality, Howson and Urbach contend, is to give up the idea of single-case probability, but it is not clear to me how one is to draw this conclusion. A more attractive response for Popper is surely to deny the assumption of determinism present in the argument. After all, any probabilities one defines for a deterministic process cannot be genuine chances. Howson and Urbach consider this response, but are wary of basing a theory of physical probability upon the metaphysical assumption of indeterminism. As they explain, 'Our objective is to find a theory of objective probability that will fit in with the practice of statistics' (1993: 341), and of course the practice of statistics will continue regardless of whether or not the world turns out to be irreducibly stochastic.

It seems to me that the view that objective probabilities are not defined for a single experiment is equivalent to saying that physical probabilities do not belong to *complete* descriptions of an experiment. (A single, concrete, instance of an experiment type defines an exact description.) Yet if probabilities do not belong to complete descriptions of experimental conditions, then surely the only alternative is to attach them to incomplete specifications. Now, if an incomplete description is needed, then question arises as to how incomplete it will be, and in which respects. In other words, one is faced with the old 'problem of the choice of reference class' (1993: 340), which Howson and Urbach are anxious to avoid.

It is quite unclear to me how an appeal to the collective is supposed to help here. The collective,[26] after all, is the very thing this incomplete description is supposed to specify! At best one might refer to some set of actual experiments, but this would surely be no better than a single trial at defining the exact conditions for an infinite sequence of trials.

To summarize, it seems that the only way to avoid the problem of the choice of reference class is to say that each experiment is performed with maximally similar initial conditions, as this leaves one with no choice to make. Of course, for a deterministic system this means that the probabilities defined will be trivial.

### *4.7.3 Howson and Urbach*

We now return to the problem of empirical access, which Howson and Urbach claim to have solved on behalf of von Mises. The basis of their position is a version of Miller's Principle, for which they offer a proof.

Their version of Miller's Principle (1993: 345), applied to the case of a coin-tossing experiment, is that $P_K(H \mid f(H) = p) = p$. In other words, if you know that a certain coin produces a collective with frequency $p$ (with respect to the property heads), and not too much else that is relevant, then your epistemic probability for heads (on any particular trial) should also be $p$. More briefly, knowledge of the frequency licenses a numerically equivalent degree of belief. Since Colin Howson is responsible for the chapter in which this principle appears, I shall refer to it as Howson's Principle.

If Howson's Principle is true, then there certainly is no problem of empirical access for von Mises. The principle provides exactly the probability measure required, in other words, to rule out deceitful collectives as massively improbable compared to their honest competitors. It says that our beliefs about finite initial segments of outcome sequences should depend strongly upon what we suppose to happen in the limit. This is counter-intuitive, in view of the fact that the limit of the sequence $f_n(H, \mathbf{s})$ is entirely independent of the first $r$ terms of **s**, in the sense that the initial segment of **s** can be altered at will without any effect on $f(H, \mathbf{s})$. We should therefore examine the argument for Howson's Principle. (I have altered the terminology of this to fit with chapter 2 of this book, but the substance is the same.)

Howson considers the situation where a coin is tossed a certain number of times to produce an outcome sequence **s**, where each term is $H$ or $T$, and a person $X$ is given an opportunity to buy the contract [\$1 if $s_1 = H$]. $X$ is given only the information that, if the tosses were continued indefinitely, the outcomes would constitute a collective with $f(H) = p$. Suppose that $X$ considers some amount \$$p'$ to be a fair price for this contract. $X$ should then also be willing to pay \$$p'$ for the contract [\$1 if $s_2 = H$], and indeed for any contract of the form [\$1 if $s_r = H$]. Let us then suppose $X$ is willing, as he should be, to buy all of these contracts. In this case, after $n$ trials in which $m$ had the outcome heads, his net gain is $m - np'$, that is, $n(m/n - p')$. Now since the limit of $m/n$ as $n$

tends to infinity is $p$, if $p' > p$ then there is some $n_0$ such that $X$ has made a loss after $n_0$ trials, and this loss is never recovered thereafter. It follows therefore that $\$p'$ is too much to pay for the contract, since it leads $X$ to a purchase that is a certain loss. By a similar argument, anything less than $\$p$ would be too little, and so $\$p$ is the fair price. Thus, by the definition of epistemic probability, $P_K(H \mid P(H) = p) = p$, which is the required result.

The argument seems to consider the case where $X$ buys all the contracts [\$1 if $s_r = H$], for every $r$, at the price $\$p'$ each. $X$ then loses, and gains, infinite amounts of money. These quantities are not comparable, so it makes no sense to ask whether $X$ loses or gains overall. In the case where $X$ buys all the contracts, therefore, he is not obviously irrational, and so the value $p'$ is not obviously incorrect. But wait a minute! Howson showed, apparently contrary to this, that after some finite number $n_0$ of trials $X$ 'goes into the red' and never comes out! This reasoning seems to indicate that the infinite losses and gains, taken together, should be considered an overall loss. How can there be this conflict?

The apparent contradiction is due to value being a cardinal quantity, whereas the limiting relative frequency depends essentially on the ordering of the outcomes. Let us suppose $p' = \frac{1}{2}$, and imagine the losses and gains to be measured in gold coins. Then each losing contract costs $X$ one coin, and each winning contract pays out one coin. It may help to think of each coin being marked with the numeral for $r$, if it is concerned with the $r^{\text{th}}$ trial, and suppose that each coin is only used once. After all is finished, therefore, there are two piles of coins, one of coins lost and the other of coins gained. The constitutions of these two piles of coins are entirely independent of how the trials are ordered. A reordering of the trials may rearrange the piles, but exactly the same coins, marked with the same numerals, will be present. Wealth is a cardinal quantity; it does not depend upon ordering.

On the other hand, a reordering of the trials may have a very large effect on $f(H)$. After reordering, $f(H)$ may have any value between 0 and 1, or indeed no value at all, as the sequence $f_n(H)$ may no longer converge. Thus, the fact that the sequence $n(m/n - p')$ goes permanently negative at some point does not mean that $X$ suffers an overall loss. By a well-chosen reordering of the trials, this sequence will instead go permanently positive at some finite point, but still the piles of coins are the same! Wealth cannot be measured by anything that is sensitive to ordering in this way.

As I describe the situation, $X$ buys all these contracts as a bundle, that is, 'all at once,' so to speak. In Howson's argument, however, it seems rather that $X$ buys them 'one at a time,' each one just before the coin is tossed, although of course $X$ is not permitted to see the outcomes. (If $X$ gets to see the outcome of each bet before deciding whether to place the next, then he has more information about the outcomes than merely that they are a collective with frequency $p$.) I do not see, however, how this difference of presentation makes any concrete difference. In both cases $X$ decides to buy all the contracts, and makes certain gains and losses.

There remains this fact that, for some finite $n_0$, $X$ will dip into the red at $n_0$ and never return to the black. Although this is a logical consequence of the fact that $f(H) = p$ – that is, it follows from a fact about the infinite tail of the outcome sequence – it looks rather like a *finite* consequence. Indeed, I think the initial plausibility of Howson's argument stems from this equivocation. The illusion can be improved if we suppose that all the coin tossing is done in advance, with the results kept secret, so that we are dealing with a particular collective **s**. There is now some definite number $k$ (let us say 865 for definiteness), such that $X$'s final slide into ruin begins at $k$. $X$ even knows that there is such a $k$, although he does not know the actual value. One may feel that it must be possible to construct some irrational bet for $X$ that is forced upon him by his being willing to pay \$$p'$ for [\$1 if $s_r = H$].

If buying all these contracts is not irrational, as the gains and losses go to infinity and become incomparable, then perhaps buying some finite initial segment is? It does not take much thought, however, to see that this is not the case. If $X$ buys all the contracts up to $r = n$ say, then whatever the value of $n$ it could still be less than $k$, for all $X$ knows. The only way to ensure that $X$ keeps buying contracts after $k$ is to make him buy them all.

It is quite clear that Howson's Principle cannot be justified using an argument of this type, so his attempt to rehabilitate von Mises's theory fails. Also, even if Howson did solve the problem of empirical access, the cardinal problem of frequency theories would remain.

### 4.7.4 Van Fraassen

Van Fraassen's frequency theory (1980: 190–6) is essentially the same that of von Mises and Popper,[27] although there are some improvements in matters of detail. He defines a *good family* as a class of collectives with certain properties. These restrictions on the class of collectives used in the definition of probability mean that the probability function defined is a countably additive measure. These refinements do not, of course, have any impact on the two problems that face frequency theories, so let us look at what van Fraassen has to say about them.

Van Fraassen does not mention the cardinal problem for frequency theories. His concern is to define a set of collectives that generate relative frequencies with the right *formal* properties. He appears not to notice that the outcome sequences excluded from the 'good family' are all quite possible, in any reasonable sense. It is hard to see what he might be thinking.

There is, however, an attempt to deal with the problem of empirical access (1980: 194–6). Van Fraassen's basic idea is to regard a finite sequence of actual outcomes as a 'random sample' from one of the collectives in the good family. He likens this random sampling to the random selection of marbles from a finite barrel where, as is well known, the proportion of marbles with a given property $A$ determines a sampling distribution for $A$. Thus, examination of a finite set of randomly selected marbles gives useful information about the proportion of $A$ in the barrel.

Van Fraassen seems willing to consider cases where the actual set of experiments is infinite, even though we only know the outcomes for a finite number of them. It seems like a good idea to discuss his response to the problem of empirical access in this context, as it means we are dealing with a particular collective **s**. For definiteness, let us suppose that we are dealing with tosses of a fair coin, so that $f(H, \mathbf{s}) = \frac{1}{2}$. Van Fraassen then regards the outcome of a single toss as a random selection from **s**. Since, in a manner of speaking, one-half of the terms in **s** are heads, it seems that the probability of heads on a single toss is also one-half.

What kind of probability is this? Is it physical or epistemic? It surely cannot be physical, for van Fraassen has already given an account of physical probability, and it did not involve the notion of random selection. So it must be epistemic probability; we have something rather like Howson's Principle, in that the epistemic probability of heads (presumably given that $f(H, \mathbf{s}) = \frac{1}{2}$) is one-half on each toss. We found that Howson's argument for his principle was fallacious, so we should see if van Fraassen has anything better.

Unfortunately there is no argument provided. Instead, van Fraassen merely tries to convey 'an intuitive idea of how the theoretically predicted frequencies are derived' (1980: 195). The intuitive idea is that we regard the collective **s** as a barrel of marbles. Now, everyone agrees that if a barrel of marbles has a known relative frequency of $\frac{1}{2}$ black marbles, then the epistemic probability of an arbitrary marble being black is also $\frac{1}{2}$. Thus, it seems, the epistemic probability that an arbitrary term in **s** is $H$ is then just the relative frequency of $H$ in **s**.

We should note first that the relative frequency of **s** depends essentially on the ordering of terms, as mentioned before. The usual idea of a 'random element' of a set $S$, in the epistemic sense, is of an object whose only *known* property is that it is a member of $S$. Thus, if a random term in **s** has a probability $\frac{1}{2}$ of being heads, then the ordering of **s** must be given. If this is so, however, then it is quite unclear how it can be shown that $P_K(s_i = H \mid f(H, \mathbf{s}) = \frac{1}{2}) = \frac{1}{2}$. In the case of a finite barrel of marbles, one uses the *symmetry* between the marbles in the barrel to argue that each marble has the same epistemic probability of being the selected one; the result then follows by the addition axiom of probability. If the ordering of **s** is known, however, as it must be for a relative frequency to be defined, then this destroys the symmetry between the terms. I have no idea how Howson's Principle can be derived along these lines, therefore. Judging from his cheerful talk of 'standard statistical calculation,' such as in the $\chi^2$ test, van Fraassen seems unaware of the theoretical difficulties involved here.

### *4.7.5 Conclusion*

It is perhaps surprising that frequency theories have such difficulty getting the right link between probability and frequency, since probability on these accounts is *defined* as frequency! On these accounts the connection between probability and frequency is both too tight and too loose at the same time – too tight for infinite outcome sequences, and too loose for finite ones. It is too tight in the infinite case, since it makes limiting relative frequencies *necessarily* equal to chances, which is overly strong. This is the cardinal problem. It is too loose in the finite case, as for a finite outcome sequence **s** the limiting relative frequency is irrelevant, and it does not provide an epistemic probability for **s**. This is the problem of empirical access.

Howson perceives that to solve the problem of empirical access one must be able to derive epistemic probabilities over finite sequences of actual outcomes from knowledge of the limiting relative frequency. Unfortunately, however, the principle required is very hard to justify, due to convergence being a tail event as well as sensitive to ordering. Limiting relative frequencies in infinite outcome sequences are used instead of real frequencies in finite outcome sequences as a 'FAPP' solution to the cardinal problem, but even if this were satisfactory in itself (which it is not), it creates another problem that is just as severe. The relative frequency is then pushed infinitely far away, out to the limit, making it empirically irrelevant.

## 4.8 CONDITIONAL CHANCES

Since, according to the causal theory, the chance function is just the logical-probability function, given certain information, it may be conditioned in the usual way. That is to say, conditioning has its usual meaning, in terms of updating beliefs in light of new information, and so the Principle of Conditioning is satisfied. In general we have $P_{\mathbf{X}}(A \mid B) = Pr(A \mid \mathcal{L}_{\mathbf{X}} \;\&\; bc_{\mathbf{X}} \;\&\; B)$.

Intuitively it appears that conditional chances are meaningful. It surely makes sense, for example, to say, 'The chance that the gun will fire, given that the trigger is pulled, is 0.97.' There even seem to be cases where one can condition the chance function upon future events; in other words, the term $P(A \mid B)$ may be meaningful even if the event $A$ occurs before event $B$.

For an example of such conditional chances, let us consider an experimental arrangement that has a radioactive source encased in such a way that particles can only escape in one direction. A short distance away there is a detector that clicks each time a particle is registered. (To keep things simple, we shall suppose that time is discrete.) The detector is imperfect, in that each time a particle arrives it has only a 95 per cent chance of firing. Also, at any time when no particle arrives, there is a 1 per cent chance of the detector firing.[28] At each time, the chance of a particle being emitted is 0.2. Times $t_1$ and $t_2$ are such that if a particle is emitted by the source at $t_1$, then it necessarily arrives at the detector at $t_2$. Let $E_1$ be the event that a particle is emitted at $t_1$, and let $F_2$ be the event that the detector fires at $t_2$, whether or not a particle was present. One would normally have no qualms about calculating the following probabilities:

$$P(E_1) = 0.2$$

$$P(F_2 \mid E_1) = 0.95$$

$$P(F_2 \mid \neg E_1) = 0.01$$

$$P(F_2) = 0.19 + 0.008 = 0.198$$

$$P(E_1 \mid F_2) = 0.19/0.198 = 0.96 \text{ (approx.)}.$$

The question is what meaning can be given to the conditional chances. Some[29] have taken the 'given,' or 'conditional upon' as primitive, and incapable of further explication, but within the causal theory of chance we can do better. As stated above, $P(E_1 \mid F_2)$ for example is defined as $Pr(E_1 \mid \mathcal{L} \text{ \& } bc \text{ \& } F_2)$. There is no reason why one should be restricted to conditioning chances on past events. This is just as well, since $P(E_1 \mid F_2)$ seems to be a meaningful term. It is the chance that a particle is present, given that the detector fires. It can even be measured empirically, if there is some way to determine when particles are emitted from the source. One simply looks at the proportion of cases, among the class of detector firings, where particles are emitted the right amount of time before that firing.

It may be wondered why we cannot define $P(A \mid B)$ simply as $P(A \text{ \& } B)/P(B)$. There are two reasons for this. First, and less importantly, this definition would be rather narrow, as it would only apply to cases where *P*(*B*) exists and is non-zero. The notion of conditional probability is valid in a much wider class of cases than this.[30] Second, and more importantly, this merely formal definition does not provide any intuitive meaning for $P(A \mid B)$, and so the definition would be useless. One could define many such symbols, such as $P(A{*}B) = P(A \vee B)/P(A)$, $P(A \# B) = P(A) + P(A \,\&\, B)$, and so on, but what for? What usually happens when such formal definitions are provided is that an intuitive meaning is *smuggled in* by the use of English words like 'given,' 'conditional upon,' and so on. This is cheating, as we require a *proof* that the formal and intuitive meanings are equivalent.

# 5

# Classical Stochastic Mechanics

As stated in chapter 1, I think an account of physical chance should prove its mettle by being useful in stochastic physics. Physicists do not often say that, in order to understand some aspect of a physical theory, one must first grasp von Mises's or Popper's interpretation of probability! An analysis of chance that does at least engage certain problems in physics is surely preferable to one that is independent of them. The remaining chapters are an attempt to show that the causal theory of chance is indeed useful, if not indispensable, in understanding some aspects of physical theory.

In this chapter I shall use the causal theory of chance to develop a very general system of mechanics. Since the formalism is fundamentally probabilistic, I call it *classical stochastic mechanics* (CSM). I use the word *classical* since, as we shall see later, it is provably inconsistent with quantum mechanics, and thus is only valid in the classical limit, that is, for 'large' systems, in some sense.

## 5.1 WHAT IS CSM GOOD FOR?

The main purpose of developing CSM in this book is to act as a foundation for the later chapters on quantum mechanics. A comparison with relativity theory may be helpful here. In one of Einstein's presentations of his theory of relativity (Einstein 1922) he begins, perhaps surprisingly, with a chapter laying out Newtonian kinematics and dynamics. Why does he do this, given that Newton's theory is already well understood? The point is that he develops the old theory in a different, and much more general, framework. Instead of starting with space and time as separate structures, for example, which effectively rules out the Lorentz transformation of coordinates, he begins with a unified structure of spacetime coordinates. When the Newtonian theory is expressed in this way, the arbitrary assumptions involved become apparent, instead of being hidden. The Galilean transform, for instance, now appears to be rather arbitrary, in acting on only the spatial coordinates. The overall effect of presenting the familiar in a novel way is to broaden the mind, and particularly the imagination, to new possibilities.

In my view, to feel at home with quantum mechanics requires a similar broadening of the imagination, the acquisition of new concepts. The easiest way to introduce these, I believe, is to follow Einstein's method, and present the familiar concepts of stochastic physics within a more general framework. We will then be in a position to see the arbitrary assumptions involved, and consider alternatives. The most difficult assumption to drop is an unconscious assumption.

The opportunity to develop a general framework for stochastic mechanics is provided by the causal theory of chance. This theory enables one to define some concepts that are usually taken to be primitive, and evaluate some claims that are normally treated as axioms. Apart from the concept of chance itself, one is able to give a detailed analysis of probabilistic independence, and also of the notion of a boundary condition. The claims that are provable include the following:

1 Causal independence implies probabilistic independence.
2 Traces of an event may succeed, but cannot precede, that event.
3 Forward, but not backward, transition probabilities are lawlike and time-independent.
4 Reichenbach's Common Cause Principle holds.
5 Entropy tends to increase.

CSM is of course related to other stochastic approaches to mechanics, and these connections will be spelled out as we go along. The basic difference, however, is that CSM is to be used for theoretical rather than practical purposes. As an analogy, consider the difference between a formal system used to construct proofs in its object language, and one used to prove mathematical results (in the metalanguage) about formal systems. The former will be more complex, containing many rules of inference to facilitate concise reasoning, as well as syntactic rules permitting the omission of brackets, and so on. A formal system used as an object about which theorems are proved, however, will be 'stripped down' as far as possible. The number of connectives, quantifiers, rules of inference, and so on will be minimized for the sake of simplicity.

## 5.2 THE LAW FUNCTION

A stochastic system, $\mathbf{X}$, is modelled by a stochastic process $\{\mathbf{X}(t)\}$ whose range is $H_{\mathbf{X}}$, the class of possible histories for **X**.[1] The proposition $\mathscr{L}_{\mathbf{X}}$, which was introduced in §3.2.2, is called the *generalized lagrangian* of **X**, and is a maximal description of **X**'s dynamical nature. A possible history is a maximal description of the actual behaviour of the system that is consistent with $\mathscr{L}_{\mathbf{X}}$, so that $\mathscr{L}_{\mathbf{X}}$ determines $H_{\mathbf{X}}$. Each possible history describes the system in the same set of times $T_{\mathbf{X}}$.[2] Possible histories of a system will be denoted by bold lower-case letters, such as $\boldsymbol{x}$. Recall from §3.3.6 that the possible history that happens to be the actual history is still an abstract object. It is not identical to the concrete history, but is the maximal actual state of affairs concerning the concrete history.

We can now restate the definition of chance from chapter 4 using the language of stochastic processes. It is that $P_{\mathbf{X}}(\mathbf{X} = \boldsymbol{x}) = Pr(\mathbf{X} = \boldsymbol{x} \mid \mathscr{L}_{\mathbf{X}} \;\&\; bc_{\mathbf{X}})$, where $\boldsymbol{x}$ is a possible history of **X** and $bc_{\mathbf{X}}$ is the boundary condition for **X**. We will usually write '$P_{\mathbf{X}}(\boldsymbol{x})$' in place of '$P_{\mathbf{X}}(\mathbf{X} = \boldsymbol{x})$' for brevity's sake.

CSM is derived from five postulates concerning the nature of real systems. The first of these is as follows.

**CSM1** The chance of a history $\boldsymbol{x}$ of **X** does not depend on the boundary condition $bc$ for **X**, but only on whether or not $\boldsymbol{x}$ satisifies $bc$. In other words, if $bc_1$ and $bc_2$ are two possible boundary conditions for **X**, and $\boldsymbol{x}$ satisfies both of them, then $Pr(\mathbf{X} = \boldsymbol{x} \mid \mathscr{L}_{\mathbf{X}} \;\&\; bc_1) = Pr(\mathbf{X} = \boldsymbol{x} \mid \mathscr{L}_{\mathbf{X}} \;\&\; bc_2)$.

CSM1 gives rise to a notion of *dynamical facility*, which is a generalization of the idea of dynamical possibility that one has in deterministic mechanics. Whereas dynamical possibility is a bivalent quantity, having only the values *possible* and *impossible*, the dynamical facility of a history make take any one of a continuous range of values. Dynamical facilities are given by a measure that I call the *law function*, which is defined as follows.

**5.2.1 Definition** The law function $L_{\mathbf{X}}$ maps each history $\boldsymbol{x}$ of **X** to the chance $\boldsymbol{x}$ has if it satisfies the boundary condition. In other words, $L_{\mathbf{X}}(\boldsymbol{x}) = Pr(\mathbf{X} = \boldsymbol{x} \mid \mathscr{L}_{\mathbf{X}} \;\&\; bc)$, where $bc$ is any boundary condition that $\boldsymbol{x}$ satisfies.

The law function is thus similar to the chance function, but there is an important difference. The law function for **X** is determined entirely by $\mathscr{L}_{\mathbf{X}}$, the generalized lagrangian for **X**, and is independent of the actual boundary condition. Consider, for instance, a history $\boldsymbol{x}$ that is inconsistent with the actual boundary condition $bc_{\mathbf{X}}$. The chance of $\boldsymbol{x}$ is zero, since from $bc_{\mathbf{X}}$ we can infer with certainty that $\boldsymbol{x}$ is not the actual history, yet the law value of $\boldsymbol{x}$ will not be zero, in general. We may express the relation between $P_{\mathbf{X}}(\boldsymbol{x})$ and $L_{\mathbf{X}}(\boldsymbol{x})$ as follows:

$$P_{\mathrm{X}}(\mathbf{x}) = \begin{cases} L_{\mathrm{X}}(\mathbf{x}) & \text{if } \mathbf{x} \text{ satisfies the boundary condition} \\ 0 & \text{otherwise.} \end{cases}$$

It is clear that $L_{\mathbf{X}}$ exists if and only if CSM1 holds, since if different boundary conditions consistent with $\boldsymbol{x}$ give rise to different values of $P_{\mathbf{X}}(\boldsymbol{x})$, then there is no such thing as the chance $\boldsymbol{x}$ has for all boundary conditions it satisfies. Why do we bother to define $L_{\mathbf{X}}$, however, and why call it the 'law' function? The first point to make, in answering these questions, is that for a deterministic system the law function is equivalent to the equation of motion. This is best explained with an example. Consider a classical particle **X** in one dimension, whose equation of motion is

$$m\frac{d^2x}{dt^2} = F(t). \qquad (1)$$

Note that CSM1 is satisfied here, as any two boundary conditions consistent with $x$ give it the same chance (either 0 or 1, depending upon whether $x$ satisfies (1)). This equation defines a law function as follows. If a history $x(t)$ satisfies (1), then $L_{\mathbf{X}}(x) = 1$; if $x$ does not satisfy (1), then $L_{\mathbf{X}}(x) = 0$. In general, $L_{\mathbf{X}}$ assigns the value 1 to each dynamically possible history, and 0 to each impossible history. A deterministic equation of motion, of course, is satisfied by a history if and only if it is dynamically possible, so that it contains the same information about the histories as $L_{\mathbf{X}}$. Equations of motion are, of course, considered to be laws.

Does a stochastic equation of motion also satisfy CSM1, and define a law function? It does, as will be explained with another example. Consider a particle $\mathbf{X}(t)$ in one-dimensional Brownian motion that is buffeted by a random force $\mathbf{F}(t)$ due to collisions with other particles. If $\alpha$ is a friction coefficient that depends on the viscosity of the fluid, and $\mathbf{V}(t)$ is the particle velocity, then the particle can be modelled by Langevin's equation:[3]

$$m\frac{d\mathbf{V}(t)}{dt} = -\alpha\mathbf{V}(t) + \mathbf{F}(t). \qquad (2)$$

What is the law value of a particular history[4] $\boldsymbol{v}(t)$, according to this equation? Let $f_v(t)$ be the forcing function that, together with $\boldsymbol{v}(t)$, satisfies equation (2).

$f_\nu(t)$, if it exists, is assigned a probability by the stochastic process $\mathbf{F}(t)$. This probability, $p$ say, is also the law value of $\nu(t)$. It is the physical chance of $\nu(t)$, if it satisfies the boundary condition. Note that the chance of $\nu(t)$ does not depend on the boundary condition as such, according to equation (2), but only on whether the boundary condition is satisfied.

Other stochastic equations of motion, such as the Fokker-Planck equation,[5] can also be expressed as a law function. These equations all have the property expressed in CSM1, that the chance of a history does not depend upon the specific boundary condition applied, but only on whether or not the boundary condition is satisfied. Applying a boundary condition to a stochastic equation of motion, one obtains a chance distribution over the class of histories, and so the stochastic differential equation does the same job as the law function. The law function gives far less insight than a differential equation into the dynamics of a system, of course, as the individual terms of an equation can be given individual physical interpretations. Since we do not consider any particular systems in this chapter, however, this is not a handicap. Moreover, the law function has the advantage that we do not have to specify the members of $H_\mathbf{X}$ – they need not even be mathematical functions.

As stated above, the law function is a measure of the 'dynamical facility' of a possible history. If a history has law value 0, for example, then it has no dynamical facility at all – the system simply cannot follow that trajectory. A law value of 1, by contrast, represents perfect dynamical facility. Once that trajectory has been embarked upon, the system will necessarily follow it all the way. In a stochastic system, the dynamical facility of a history will lie somewhere in between these extremes.

It is useful to have a measure that is determined by $\mathcal{L}_\mathbf{X}$ alone. Many properties of stochastic systems are independent of the boundary condition. These properties correspond to rather simple arithmetical relations between law values, and more complex ones between chances, as one would expect. We shall see, for example, that the Markov property can be expressed as a very simple constraint on $L$.

In the above discussion I have assumed that the law function is defined on individual histories, but in general, like any measure, it is defined on classes of histories. This is not a serious complication, as we can modify Definition 5.2.1 so that $L_\mathbf{X}(C) = Pr(\mathbf{X} \in C \mid \mathcal{L}_\mathbf{X} \;\&\; bc)$, where $C$ is a class of histories, each of which satisfies $bc$. This definition specifies $L_\mathbf{X}$ on an algebra of rectangles, and then we use the Carathéodory technique[6] to extend $L_\mathbf{X}$ to a $\sigma$-algebra. To specify the details of this construction would require some specific assumptions about $H_\mathbf{X}$, the class of possible histories. In a theoretical work such detail is surely unnecessary. Since the existence of $L_\mathbf{X}$ on a $\sigma$-field of subsets of $H_\mathbf{X}$

presents no more difficulty than the existence of a normalized measure such as $P_{\mathbf{X}}$, there is no reason to suppose that any problem will arise here.

## 5.3 RELEVANCE AND CORRELATION

Before we can consider the dynamics of composite systems, we must understand some basic facts about relevance and correlation. This work is foundational to the major problem investigated in this chapter and the next, namely, how to understand correlation with respect to chance. In this section we shall define all the necessary concepts, lay down some postulates, and prove some useful theorems.

According to the causal theory of chance, relevance and correlation are, at bottom, the same relation. This is counter-intuitive, since 'relevance' is associated with epistemology and 'correlation' with physics, but will be argued for. Let us begin by defining epistemic relevance in the standard way.

**5.3.1 Definitions** (i) $A$ is relevant to $B$ within epistemic state $K$ iff $P_K(B \mid A) \neq P_K(B)$.
(ii) $A$ is positively (negatively) relevant to $B$ iff $P_K(B \mid A) > (<)\ P_K(B)$.

For the first definition, the probabilities are intervals in general, and single values only as a special case. Thus, for example, if $P_K(B \mid A) = [0.1, 0.4]$ and $P_K(B) = [0.1, 0.3)$, then $A$ is relevant to $B$. Intuitively, $A$ is relevant to $B$ within $K$ just in case, for someone with epistemic state $K$, learning that $A$ is true makes some kind of difference to their attitude to $B$.

**5.3.2 Theorems** (i) Relevance is a symmetric relation.
(ii) If $A$, $B$ are mutually irrelevant, then $P_K(A \ \&\ B) = P_K(A)P_K(B)$.

The proofs of these theorems are trivial. Mutually irrelevant propositions are sometimes described as 'independent.' To avoid confusion, however, I shall reserve the term 'independent' for *causal* independence, that is, non-interaction.

The term 'correlation,' although it is usually reserved for physical contexts, has a purely formal definition. In fact, the definition of correlation is formally identical to the one given above for relevance. The difference is that the term 'correlation' is rarely (if ever) used where the probability function in question is an epistemic probability. For a correlation, the normalized measure must be either physical chance or relative frequency. There are thus two types of correlation, which we shall call *correlation with respect to chance* and *frequency correlation*. These two are quite distinct, although one may be evidence for the other. Frequency correlations are the kind that can be measured directly, by counting. They hold, of course, not between singular propositions like 'Jim is a surgeon' and 'Jim is wealthy,' but rather between properties, such as being a surgeon and being wealthy.

Frequency correlations are often good evidence for a correlation with respect to chance. Suppose there is a random-number generator that produces numbers between 1 and 20, inclusively. After (say) a hundred numbers have been picked, we observe that there is a positive frequency correlation between a number thus produced being even and its being greater than 10. That is, the proportion of even numbers among those greater than 10 is higher than their proportion among numbers less than 10. This does not guarantee, but is some evidence for, a correlation with respect to chance for these two properties. The correlation with respect to chance, however, applies at the level of each individual trial rather than the reference class of 100 trials.

The frequency correlation does not entail a correlation with respect to chance, since the absence of a correlation with respect to chance is perfectly consistent with there being a frequency correlation – the frequency correlation may be, as one says, *due to chance*. See how unfortunate that expression is, from my point of view! A frequency correlation that is 'due merely to chance' is precisely one that is *not* associated with any correlation with respect to chance! I will sometimes use the term *chance correlation* as a synonym for correlation with respect to chance, for convenience. I hope that this will not cause any confusion of the kind discussed.

When is one proposition relevant to another? We will not answer this question in general, but only consider the case of two propositions that are about separate systems, which we call **X** and **Y**. **X** and **Y** are separate in the sense that they are entirely disjoint, having no common part, yet they need not be causally isolated from one another. The Earth and the Sun, for instance, are separate systems, yet they interact in various ways.

In order to talk of two systems **X** and **Y** being correlated, with respect to chance, we need to compare the chance functions for **X** and for **Y** with some sort of *joint* chance function for **X** and **Y** together. This joint chance function is most naturally and conveniently identified with the chance function of a third system, namely, the *composite* system that consists of **X** and **Y** together (and nothing else). We thus require that, for every pair of system **X** and **Y**, there exists a composite system ⟨**X**, **Y**⟩. The composite system need not have any essential unity or cohesion – it may be an arbitrary collection of spatially distant systems. It is designated a 'system' simply because it may be treated as a system in the formalism, that is, it has its own lagrangian and boundary condition. Note also that ⟨**X**, **Y**⟩ is a concrete object, and thus quite distinct from the class {**X**, **Y**}. There is no difference, for example, between ⟨⟨**W**, **X**⟩, **Y**⟩ and ⟨**W**, ⟨**X**, **Y**⟩⟩. The second postulate of CSM is therefore as follows.

**CSM2** If **X** and **Y** are systems, then so is the collection of **X** and **Y**, which we write ⟨**X**, **Y**⟩. Thus ⟨**X**, **Y**⟩ has its own lagrangian and boundary condition.

If we call the collection of all physical systems the *cosmos*, then it should be noted that CSM2 entails that the cosmos is a physical system. At this stage, in order to keep things as general as possible, we deliberately do not specify which objects count as systems in the first place.

The basic postulate concerning relevance is as follows:

**5.3.3 Postulate** If **X** and **Y** are separate systems, and $A_{\mathbf{X}}$, $A_{\mathbf{Y}}$ are propositions about **X** and **Y** respectively, then $A_{\mathbf{X}}$, $A_{\mathbf{Y}}$ are mutually irrelevant within $K_O$.[7]

As noted in chapter 2, terms like $Pr(A \mid O)$ are generally undefined unless either $A$ or $\neg A$ is a necessary truth. That is, there is no *single value* that is the correct degree of belief in $A$, determined by rationality alone, but there is always an *interval* of permitted degrees of belief, even if it is only (0,1). Moreover, according to Definition 5.3.1, relevance is defined in terms of interval probabilities, with single-point probabilities merely as a special case. $A_{\mathbf{X}}$ and $A_{\mathbf{Y}}$ are mutually irrelevant within $K_O$ just in case $Pr(A_{\mathbf{X}} \mid O)$ and $Pr(A_{\mathbf{X}} \mid A_{\mathbf{Y}})$ are the same interval.

Postulate 5.3.3 entails that, if $A_{\mathbf{X}}$ is relevant to $A_{\mathbf{Y}}$ within some epistemic state $K$, then $K$ contains some knowledge of the systems **X** and **Y**. The question now before us is, What general character must $K$ have in order for $A_{\mathbf{X}}$ to be relevant to $A_{\mathbf{Y}}$? The following theorem helps us to answer this question.

**5.3.4 Theorem** If $K$ contains knowledge of **X** only, then $A_{\mathbf{X}}$ is irrelevant to $A_{\mathbf{Y}}$ within $K$.

*Proof*: Suppose $K$ consists only of the proposition $B_{\mathbf{X}}$ concerning **X**. Then $P_K(A_{\mathbf{Y}} \mid A_{\mathbf{X}}) = Pr(A_{\mathbf{Y}} \mid A_{\mathbf{X}} \mathbin{\&} B_{\mathbf{X}})$. But $A_{\mathbf{X}} \mathbin{\&} B_{\mathbf{X}}$ is a proposition about **X** alone, and thus by Postulate 5.3.3 we have that $Pr(A_{\mathbf{Y}} \mid A_{\mathbf{X}} \mathbin{\&} B_{\mathbf{X}}) = Pr(A_{\mathbf{Y}} \mid O)$. Moreover, by the same reasoning, $P_K(A_{\mathbf{Y}}) = Pr(A_{\mathbf{Y}} \mid O)$, and so we have $P_K(A_{\mathbf{Y}} \mid A_{\mathbf{X}}) = P_K(A_{\mathbf{Y}})$, which is the result.■

By symmetry it is clear that, for $A_{\mathbf{X}}$ to be relevant to $A_{\mathbf{Y}}$ within $K$, $K$ must not merely contain information about **Y**. We see therefore that $K$ must contain information about both **X** and **Y**. This information cannot be in the form of two separate propositions, however, one about **X** and the other about **Y**, as the following theorem shows.

**5.3.5 Theorem** If $K$ consists of $B_{\mathbf{X}}$ and $B_{\mathbf{Y}}$, concerning **X** and **Y** respectively, then $A_{\mathbf{X}}$ is irrelevant to $A_{\mathbf{Y}}$ within $K$.

*Proof*:

$$\begin{aligned}
P_K(A_{\mathbf{Y}} \mid A_{\mathbf{X}}) &= Pr(A_{\mathbf{Y}} \mid A_{\mathbf{X}} \,\&\, (B_{\mathbf{X}} \,\&\, B_{\mathbf{Y}})) \\
&= Pr(A_{\mathbf{Y}} \mid (A_{\mathbf{X}} \,\&\, B_{\mathbf{X}}) \,\&\, B_{\mathbf{Y}}) \\
&= R(A_{\mathbf{Y}} \,\&\, B_{\mathbf{Y}} \mid A_{\mathbf{X}} \,\&\, B_{\mathbf{X}},\, B_{\mathbf{Y}} \mid A_{\mathbf{X}} \,\&\, B_{\mathbf{X}}) \\
&= R(A_{\mathbf{Y}} \,\&\, B_{\mathbf{Y}},\, B_{\mathbf{Y}}).
\end{aligned}$$

$$\begin{aligned}
\text{Now, } P_K(A_{\mathbf{Y}}) &= Pr(A_{\mathbf{Y}} \mid B_{\mathbf{X}} \,\&\, B_{\mathbf{Y}}) \\
&= R(A_{\mathbf{Y}} \,\&\, B_{\mathbf{Y}} \mid B_{\mathbf{X}},\, B_{\mathbf{Y}} \mid B_{\mathbf{X}}) \\
&= R(A_{\mathbf{Y}} \,\&\, B_{\mathbf{Y}},\, B_{\mathbf{Y}}).
\end{aligned}$$

Thus $P_K(A_{\mathbf{Y}} \mid A_{\mathbf{X}}) = P_K(A_{\mathbf{Y}})$, as required.[8] ■

**5.3.6 Definition** Let $B_{\langle \mathbf{X},\mathbf{Y}\rangle}$ be a proposition about $\langle \mathbf{X}, \mathbf{Y}\rangle$. Then $\{B_{\langle \mathbf{X},\mathbf{Y}\rangle}\}_{\mathbf{X}}$ is the maximal proposition about **X** that is logically entailed by $B_{\langle \mathbf{X},\mathbf{Y}\rangle}$.

**5.3.7 Definition** A proposition $B_{\langle \mathbf{X},\mathbf{Y}\rangle}$ concerning **X** and **Y** is said to be *factorizable* just in case it is equivalent to the conjunction $B_{\mathbf{X}}$ & $B_{\mathbf{Y}}$, where $B_{\mathbf{X}} = \{B_{\langle \mathbf{X},\mathbf{Y}\rangle}\}_{\mathbf{X}}$ and $B_{\mathbf{Y}} = \{B_{\langle \mathbf{X},\mathbf{Y}\rangle}\}_{\mathbf{Y}}$.

The propositions $B_{\mathbf{X}}$ and $B_{\mathbf{Y}}$ may be considered 'factors' of $B_{\langle \mathbf{X},\mathbf{Y}\rangle}$, since truth-functional conjunction is sometimes called the logical product. Indeed, if we represent The True by 1 and The False by 0, then conjunction is multiplication. Note that $B_{\langle \mathbf{X},\mathbf{Y}\rangle}$ is only considered factorizable if each of its factors is only about one system.

**5.3.8 Corollary** If $K$ consists of $B_{\langle \mathbf{X},\mathbf{Y}\rangle}$ and $A_{\mathbf{X}}$ is relevant to $A_{\mathbf{Y}}$ within $K$, then $B_{\langle \mathbf{X},\mathbf{Y}\rangle}$ is not factorizable.

*Proof*: Immediate from Theorem 5.3.5.■

What are some examples of non-factorizable propositions about **X** and **Y**? Conditionals are perhaps the most obvious kind. The statement 'If **X** = 1 then **Y** = 1,' for instance, cannot be factorized. The same holds for disjunctions, of course. A more interesting class of examples is where *K* contains information about causal interaction, such as '**X** interacts with **Y**.' It is clear that this cannot be factorized (it is not equivalent to, for instance, '**X** interacts with something and **Y** interacts with something').

If an epistemic state *K* does not factorize with respect to the two systems **X** and **Y** that it concerns, then we might say that the knowledge *K* of the two systems is *entangled*.[9] Alternatively, we might say that the knowledge *K* is *non-local*, as it irreducibly concerns both systems together, which may be far apart in spacetime.

It is important to realize that the converse of Corollary 5.3.8 does not hold, as not every non-factorizable proposition gives rise to relevance. Consider, for instance, the information that **X** does not interact with **Y**. This cannot be factorized, yet intuitively at least it does not create an inferential 'bridge' between **X** and **Y**; that is, knowing that **X** and **Y** do not interact does not enable one to make inferences about **Y** from statements about **X**. The presence of non-factorizable knowledge is a necessary condition for relevance, but not a sufficient condition. I shall assume, however, that if the epistemic state *K* says an **X**–**Y** interaction *does* (or might) occur, then this does provide an inferential bridge between **X** and **Y**.

## 5.4 CHANCE IN A COMPOSITE SYSTEM

Let **Z** be the composite system ⟨**X**, **Y**⟩. Since **Z** is a system, it must have a lagrangian $\mathcal{L}_Z$, which determines a class of histories $H_Z$. The first question to examine, therefore, is how $H_Z$ is related to $H_X$ and $H_Y$. The usual assumption to make here is that each member of $H_Z$ can be identified with a pair of histories, one history for **X** and one for **Y**. In other words, if $z \in H_Z$, then there exist $x \in H_X$ and $y \in H_Y$ such that $z = \langle x, y \rangle$. Since $x$, $y$, and $z$ are states of affairs, we can also write $z = x \,\&\, y$. Using the notation of Definition 5.3.6, we shall say that $z$ *factorizes* into $x \,\&\, y$. Also note that, according to this postulate, $H_Z$ is the cartesian product of $H_X$ and $H_Y$.

This assumption that histories of composite systems factorize seems intuitively correct, and also has powerful consequences for the structure of the chance function. I do not accept it as generally valid, however. In §3.3.7 I claimed that a pair of interacting stochastic systems **X** and **Y** might only have a joint history, so that neither **X** nor **Y** has its own history. In spite of this, I shall include the factorizability of histories among the postulates of CSM. While it is not universally valid, in my view, it is a defining assumption (perhaps even *the* defining assumption) of classical physics. It is needed to derive the familiar, common-sense independence properties of the chance function.

**CSM3** If **Z** is the composite system ⟨**X**, **Y**⟩, then each history $z$ of **Z** is equal to some $\boldsymbol{x}$ & $\boldsymbol{y}$, where $\boldsymbol{x}$ and $\boldsymbol{y}$ are histories of **X** and **Y** respectively.

CSM3 gives us an extra theorem about relevance, that inferential bridges can be destroyed by adding extra information. Suppose the proposition $B_{\langle\mathbf{X},\mathbf{Y}\rangle}$ about the history of ⟨**X**, **Y**⟩ is not factorizable, and makes **X** relevant to **Y**. Then consider the proposition $C_{\mathbf{X}}$ , which is a *complete* description of **X**'s history. If $K$ is the minimal state including $B_{\langle\mathbf{X},\mathbf{Y}\rangle}$ & $C_{\mathbf{X}}$, then is **X** relevant to **Y** within $K$? In fact it is not relevant, as $B_{\langle\mathbf{X},\mathbf{Y}\rangle}$ & $C_{\mathbf{X}}$ is factorizable, as is shown below.

**5.4.1 Lemma** If $B_{\langle\mathbf{X},\mathbf{Y}\rangle}$ is complete concerning **X**, that is, $\{B_{\langle\mathbf{X},\mathbf{Y}\rangle}\}_{\mathbf{X}}$ is complete, then $B_{\langle\mathbf{X},\mathbf{Y}\rangle}$ factorizes.

*Proof*: Let $B_{\mathbf{X}} = \{B_{\langle\mathbf{X},\mathbf{Y}\rangle}\}_{\mathbf{X}}$. By hypothesis, $B_{\mathbf{X}}$ is complete, that is, equivalent to some single history $\boldsymbol{x}$. By contrast, $B_{\langle\mathbf{X},\mathbf{Y}\rangle}$ is a disjunction of histories $z_i$, namely:

$$B_{\langle\mathbf{X},\mathbf{Y}\rangle} = \bigcup_i z_i$$

$$= \bigcup_i (\boldsymbol{x}_i \;\&\; \boldsymbol{y}_i),$$

where the union symbol denotes truth-functional disjunction. Now, since $B_{\mathbf{X}} = \boldsymbol{x}$, the $\boldsymbol{x}_i$ in the summation must all be equal to $\boldsymbol{x}$. Thus:

$$B_{\langle\mathbf{X},\mathbf{Y}\rangle} = \bigcup_i (\boldsymbol{x} \;\&\; \boldsymbol{y}_i)$$

$$= \boldsymbol{x} \;\&\; \bigcup_i \boldsymbol{y}_i$$

$$= B_{\mathbf{X}} \;\&\; B_{\mathbf{Y}}.$$

And so $B_{\langle\mathbf{X},\mathbf{Y}\rangle}$ factorizes into $B_{\mathbf{X}}$ & $B_{\mathbf{Y}}$, as required.■

**5.4.2 Corollary** If $C_X$ is complete, then $C_X$ & $B_{\langle X,Y \rangle}$ factorizes.

*Proof*: Since $C_X$ is complete, it follows that $\{C_X \ \& \ B_{\langle X,Y \rangle}\}_X$ is complete. The result then follows using Theorem 5.4.1.■

We should now examine how $L_Z$ is related to $L_X$ and $L_Y$. By the definition of $L_Z(z)$ we have

$L_Z(z) = Pr(\mathbf{Z} = z \mid \mathcal{L}_Z \ \& \ bc_Z)$, where $bc_Z$ is any boundary condition consistent with $z$.

Now, according to CSM3, we may replace $\mathbf{Z} = z$ with $\mathbf{X} = \boldsymbol{x}$ & $\mathbf{Y} = \boldsymbol{y}$, which yields:

$$\begin{aligned} L_Z(z) &= Pr(\mathbf{X} = \boldsymbol{x} \ \& \ \mathbf{Y} = \boldsymbol{y} \mid \mathcal{L}_Z \ \& \ bc_Z) \\ &= Pr(\mathbf{X} = \boldsymbol{x} \mid \mathcal{L}_Z \ \& \ bc_Z) Pr(\mathbf{Y} = \boldsymbol{y} \mid \mathcal{L}_Z \ \& \ bc_Z \ \& \ \mathbf{X} = \boldsymbol{x}) \\ &= L_Z(\boldsymbol{x}) L_Z(\boldsymbol{y} \mid \boldsymbol{x}). \end{aligned}$$

To dissect this expression further, we must consider how $\mathcal{L}_Z$ is related to $\mathcal{L}_X$ and $\mathcal{L}_Y$. We should first note that $\mathcal{L}_Z$ does not determine either $\mathcal{L}_X$ or $\mathcal{L}_Y$ (we might say that $\mathcal{L}_Z$ does not 'know' what $\mathcal{L}_X$ and $\mathcal{L}_Y$ are). This is because $\mathcal{L}_Z$ does not determine the actual history of either **X** or **Y** and, on account of the **X**–**Y** interaction, $\mathcal{L}_X$ depends on the actual history of **Y**, and $\mathcal{L}_Y$ depends on the actual history of **X**. Thus, $\mathcal{L}_Z$ together with $\mathbf{X} = \boldsymbol{x}$ determines $\mathcal{L}_Y$, but $\mathcal{L}_Z$ alone does not.

Let us now consider the special, and very important, case where **X** and **Y** cannot interact. This means that the chance of interaction is zero, so that $\mathcal{L}_Z$ & $bc_Z$ together entail that no interaction occurs. In this case, since $\mathcal{L}_Y$ does not depend on the actual history of **X**, $\mathcal{L}_Z$ (together with $bc_Z$) does know $\mathcal{L}_Y$, as well as $\mathcal{L}_X$ of course. When **X** and **Y** cannot interact, therefore, $\mathcal{L}_Z$ & $bc_Z$ can be written as $\mathcal{L}_X$ & $\mathcal{L}_Y$ & $bc_Z$. The two propositions are logically equivalent. In short, if **X** and **Y** cannot interact (for a given $bc_Z$), then, within $bc_Z$, $\mathcal{L}_Z$ factorizes into $\mathcal{L}_X$ & $\mathcal{L}_Y$. Causal independence entails that the joint lagrangian factorizes.

Another important question is how $bc_Z$, the boundary condition for **Z**, relates to $bc_X$ and $bc_Y$. In the spirit of CSM3, I shall assume that $bc_Z = bc_X$ & $bc_Y$, regardless of whether **X** and **Y** may interact or not. I shall later argue that $bc_Z$, $bc_X$, and $bc_Y$ are states, and that states are just very short parts of a history, so this assumption is really a consequence of CSM3.

**5.4.3 Theorem** If, for a particular $bc_Z$, systems **X** and **Y** cannot interact with each other, then they are uncorrelated w.r.t. chance, that is, $P_Z(x \,\&\, y) = P_X(x)P_Y(y)$.

*Proof:*

1. $P_Z(x \,\&\, y) = Pr(x \,\&\, y \mid \mathcal{L}_Z \,\&\, bc_Z)$
2. $= Pr(x \,\&\, y \mid \mathcal{L}_X \,\&\, \mathcal{L}_Y \,\&\, bc_X \,\&\, bc_Y)$
3. $= Pr(x \mid \mathcal{L}_X \,\&\, \mathcal{L}_Y \,\&\, bc_X \,\&\, bc_Y)Pr(y \mid x \,\&\, \mathcal{L}_X \,\&\, \mathcal{L}_Y \,\&\, bc_X \,\&\, bc_Y)$
4. $= Pr(x \mid \mathcal{L}_X \,\&\, bc_X)Pr(y \mid \mathcal{L}_Y \,\&\, bc_Y)$
5. $= P_X(x)P_Y(y)$

Line 1 follows from the definition of $P_Z$. Line 2 just factorizes $\mathcal{L}_Z \,\&\, bc_Z$. Line 3 applies Axiom 4, the Principle of Conditioning. Line 4 applies Theorem 5.3.5, since $x \,\&\, \mathcal{L}_X \,\&\, bc_X$ concerns **X** only, and $\mathcal{L}_Y \,\&\, bc_Y$ concerns **Y** only. Line 5 follows from line 4, using the definitions of $P_X$ and $P_Y$.■

## 5.5 SUB-HISTORIES, STATES, AND MARKOV SYSTEMS

So far in this chapter we have only considered one kind of event within a system **X**, namely, an entire history for **X**. Physicists frequently refer to events that are less extensive than whole histories, however, so we shall look at what these 'smaller' events may be. It will then be possible to define and discuss the Markov property of stochastic processes.

Consider a system **X** that exists in a time interval $T_X$. This interval can be mentally split into sub-intervals, of course. Such a division of $T_X$ automatically splits up $x$, the actual history of **X**, into corresponding chunks. For instance, if $T_X = [0,10]$, and we split it into $[0,4)$ and $[4,10]$; then we can define $x_1$ as the part of $x$ that occurs in $[0,4)$, and $x_2$ as the part that exists in $[4,10]$. We note that the motion or change that occurs in $x$ also occurs in $x_1$ and $x_2$ together: some of the motion is in $x_1$, and the rest shows up in $x_2$. If we partition $T_X$ into finer time slices, such as $[0, \delta)$, $[\delta, 2\delta)$, ..., $[10 - \delta, 10]$, where $\delta > 0$, then the amount of change in each sub-interval is reduced, but it still appears that all the change in $x$ shows up in the time slices.

So far I have not introduced the concept of a *state* of a system, which is one important kind of event, but it seems to be closely connected to the idea of a time slice. Perhaps a possible state of **X** just is a thin time slice of a possible history of **X**, such as the piece of $\boldsymbol{x}$ in $[n\delta, (n + 1)\delta)$? This does not seem quite right, however, since regardless of how small $\delta$ may be, **X** undergoes some change in this sub-history. States are supposed to be 'static,' and not containing any motion at all. To get sub-histories that are also states, therefore, perhaps we can partition $T_{\mathbf{X}}$ into the individual real numbers of which it is composed. The sub-histories of **X** corresponding to this partition will be very thin indeed, and it does appear that each one contains no change.

Unfortunately, we run into Zeno's paradox of the arrow here. If each of these very thin sub-histories contains no motion at all, then where is all the change in $\boldsymbol{x}$? Consider, for instance, the flight of an arrow from the bow to the target, a distance of perhaps 100 metres. The total history contains 100 m of motion, and so the collection of sub-histories should contain 100 m of motion as well. This is the case for a finite partition into sub-histories, but not for an infinite partition into states. In each state the arrow moves zero metres, and an infinite sum of zeros is still zero. It appears that all the motion occurs 'between' the states, but in that case we do not have a genuine partition of $T_{\mathbf{X}}$, as something in the history $\boldsymbol{x}$ has been lost.

For this reason it seems to me that the concept of a state is something of a mathematical fiction, rather like the Dirac delta function. It should not be taken literally as a real entity, but rather as the fictional limit of a real sequence. In the limit as a (finite) time slice tends to a point, the amount of change may be treated as negligible, and so that time slice can be regarded as a state. This is essentially the same problem that we face in the classical mechanics of a single particle. Although the whole trajectory is captured by the position function $\mathbf{r}(t)$, the state at a single time $t'$ is not fully represented by $\mathbf{r}(t')$, as this gives no information about the momentum of the particle. To fix the momentum at $t'$ we require a short (but finite) portion of the trajectory that includes $t'$. Since this time interval may be arbitrarily short, it makes sense to regard a state as a limiting concept, as in the following definition.

**5.5.1 Definition** A possible *state* of a system **X** at time $t$ is the limit of a sequence of time slices of a possible history of **X**, which converge to the interval $\{t\}$.

We must now consider some general properties of stochastic systems in relation to time. The main such property is known as the Markov condition, which real systems are frequently assumed to possess, as least to a good level of approximation. We shall now examine the metaphysical basis of the Markov condition by attempting to derive it from more fundamental assumptions.

The Markov condition may be expressed in various equivalent ways, one of which is as follows:

**5.5.2 Definition** A system **X** is Markov iff the distributions for sub-histories in disjoint intervals are conditionally irrelevant, given a sub-history in between them.

One consequence of the Markov property is that if $t_0 < t_1 < t_2$, and you want to predict the behaviour of a system at $t_2$, then knowledge of the state of the system at $t_1$ renders the state at $t_0$ irrelevant. This is sometimes expressed by saying that the state at $t_1$ *screens off* the state at $t_0$, as far as the state at $t_2$ is concerned. It seems clear that, if this is true, then the system at $t_0$ cannot directly interact with the system at $t_2$, under the general principle that causal interaction implies logical relevance. To make sense of the idea of different time slices of a system interacting with each other, we need to consider the finite time slices of a system as systems themselves. We therefore require the following postulate.

**CSM4** Finite time slices of a system are causally isolated systems. (Each time slice has its own lagrangian and boundary condition, and non-intersecting time slices of a given system cannot interact with each other.)

If CSM4 is true, then we may be able to derive an expression relating the chance distributions of the whole system to the chance distributions of the time slices, which would be analogous to Theorem 5.4.1. One problem that arises here is the question of boundary conditions for the time slices. We have not yet discussed what form of constraint the boundary condition is in general, and for this reason it is hard to make sense of CSM4. If a time slice is an autonomous system then it should not only have its own boundary condition, but also the same basic kind of boundary condition as the total system. As we shall see, this means that CSM4 is quite restrictive, as it rules out most types of boundary condition.

## 5.6 BOUNDARY CONDITIONS AND TIME

In deterministic mechanics the topic of the metaphysical status of boundary conditions is not often discussed. The behaviour of a system is often thought to be produced by the dynamical properties of the system alone, with the boundary condition seen as mere 'information,' rather than as a second cause.[10] I believe that this point of view results from the fact that, in a deterministic system, the boundary condition is 'invisible,' in a sense that is explained below. In a stochastic system, by contrast, the boundary condition is 'visible' and so has to be taken seriously as a physical cause.

In a deterministic system $\mathbf{X}$ there are only two grades of dynamical facility, namely, perfect facility and none at all, which correspond to the two law values 1 and 0 respectively. One consequence of this fact is that a given chance function over $H_\mathbf{X}$ can be generated by many different boundary conditions. Consider, for instance, the boundary condition $\mathbf{X}(t_1) = x_1$. This fact, together with $\mathcal{L}_\mathbf{X}$, entails that the actual history is the unique $\boldsymbol{x} \in H_\mathbf{X}$ such that $\boldsymbol{x}(t_1) = x_1$ and $L_\mathbf{X}(\boldsymbol{x}) = 1$. But now, any other complete boundary condition consistent with $\boldsymbol{x}$, such as $\mathbf{X}(t_2) = \boldsymbol{x}(t_2)$, will yield that very same history $\boldsymbol{x}$. Thus, if one only has the chance function $P_\mathbf{X}$ for a deterministic system $\mathbf{X}$, then one has no idea which of these possible boundary conditions is the real one – it could be set at any time in $T_\mathbf{X}$. In this sense, the boundary condition of a deterministic system, considered as a cause, is invisible (to the chance function).

More precisely, every deterministic system has the following three properties:

1 The nature of the real boundary condition is empirically inaccessible. (One cannot infer anything about the boundary condition from observation or experiment.)
2 One cannot explain any observed phenomenon (natural or experimental) by a hypothesis about the boundary condition.
3 In particular, one cannot explain the arrow-of-time phenomena by a hypothesis about the boundary condition.

To see that (1) is true, consider two rival hypotheses $h_1$ and $h_2$ about the boundary condition, where $h_1$ says that the boundary condition is $bc_1$ and $h_2$ says that it is $bc_2$. If $\boldsymbol{x}$ is the observed actual history for $\mathbf{X}$, and $h_1$, $h_2$ are consistent with $\boldsymbol{x}$, then $P_\mathbf{X}(\boldsymbol{x} \mid \mathcal{L}_\mathbf{X} \,\&\, bc_1) = P_\mathbf{X}(\boldsymbol{x} \mid \mathcal{L}_\mathbf{X} \,\&\, bc_2)$. Then, by Miller's principle, we have that $P_K(\boldsymbol{x} \mid h_1) = P_K(\boldsymbol{x} \mid h_2)$, so that

$$\frac{P_K(h_1 \mid \mathbf{x})}{P_K(h_2 \mid \mathbf{x})} = \frac{P_K(\mathbf{x} \mid h_1) P_K(h_1)}{P_K(\mathbf{x} \mid h_2) P_K(h_2)}$$

$$= \frac{P_K(h_1)}{P_K(h_2)}.$$

Thus, the empirical datum $\boldsymbol{x}$ has no bearing on the relative probability of $h_1$ and $h_2$.

To see that (2) holds, recall from §4.6 that to explain a phenomenon is to infer it from a hypothesis about its cause. Thus, to explain $\boldsymbol{x}$ by $h_1$ one must infer $\boldsymbol{x}$ from $h_1$. One problem here is that there are too many rival explanations. For instance, one could explain the same datum $\boldsymbol{x}$ by the alternative hypothesis $h_2$. Since every boundary condition consistent with $\boldsymbol{x}$ successfully entails $\boldsymbol{x}$, $\boldsymbol{x}$ itself does not help us to choose between these competing explanations. Another problem is that even the most bizarre counterfactual data could be explained just as easily as the actual data. As long as a history is dynamically possible, it will be actual provided it satisfies the boundary condition.

To explain phenomena within a deterministic system by a hypothesis about the boundary condition is thus a very weak explanation. The choice of explanatory hypothesis is arbitrary, and also by this method we could explain just about anything. Such explanatory power is not considered a theoretical virtue!

It may be possible to judge some hypotheses to be *a priori* more (epistemically) probable than others but, since the boundary condition is an external constraint on the cosmos, this means that the explanation is more metaphysical than physical. While I do not wish to rule such explanations out as generally invalid, I take it that a physical explanation, if available, is preferable to a metaphysical one.

Claim (3) concerns the famous 'arrows of time,' which are discussed in §5.7. They are temporally asymmetric physical phenomena that are awkward to explain, in light of the fact that the laws of physics seem to be time symmetric. Claim (3) follows from (2), of course, but some more particular remarks may be made. It is orthodox among physicists to attribute temporal asymmetries to special initial conditions, as with the following authors:[11]

> While matters are by no means universally agreed upon, the most plausible view at the present time seems to be that in order to get a reasonable picture of the entropic increase accompanying expansion of our current phase of (at least the 'local') universe, we must impose a low entropy initial condition on the big-bang singularity.

> [W]e are led more or less inevitably to cosmological considerations of an initial 'state of the universe' having a very small Boltzmann entropy. That is, the universe is pictured to be born in an initial macrostate $M_0$ for which [the phase space volume] is a very small fraction of the 'total available' phase space volume.

> [I]t is necessary to add to the physical laws the hypothesis that in the past the universe was more ordered, in the technical sense, than it is today ... to make an understanding of the irreversibility.

For such an explanation to work one must assert that some hypotheses about the boundary condition have much higher epistemic probabilities than others, which is a metaphysical rather than a physical statement. If this were the only explanation on offer then we would perhaps be justified in accepting it. I will argue, however, that a physical explanation is available, within stochastic physics.

To sum up, it seems that physics requires the concept of probability. If one accepts indeterminism then these probabilities exist within the cosmos, so to speak, and are thus squarely within the domain of physics. If one insists on determinism, on the other hand, then the probabilities are forced out of the physical cosmos itself and into the Great Beyond.

For a stochastic system, none of (1)–(3) holds, as we shall now see. Consider, for instance, the two boundary conditions $\mathbf{X}(t_1) = x_1$ and $\mathbf{X}(t_2) = x_2$ for the stochastic system $\mathbf{X}$. Do they give rise to the same chance function $P_{\mathbf{X}}$? They do not, as $P_{\mathbf{X}}(\mathbf{X}(t_1) = x) = 1$ in the first case, but in the second case it will have some value less than one. If $t_1$ and $t_2$ are widely separated in time, then $P_{\mathbf{X}}(\mathbf{X}(t_1) = x_1)$ may be quite low for the boundary condition $\mathbf{X}(t_2) = x_2$.

Thus, if $\mathbf{X}$ is stochastic, then $P_{\mathbf{X}}$ depends strongly on the time at which the boundary condition is set. Now, as shown in chapter 4, $P_{\mathbf{X}}$ is empirically accessible by measuring frequencies, so it follows that facts about the boundary condition are also empirically accessible, at least to some extent. Indeed, in this section I will argue that from certain data we can infer that the boundary condition for the cosmos is set at the Big Bang.

We should also note that, for a stochastic system, a boundary condition that physically constrains the actual history at a single point in time introduces a temporal asymmetry, and a local direction of time. (One temporal direction points towards the time of the boundary condition, and the other points away from it.) This is particularly clear in the case where the boundary condition is set at one extremum of the time interval $T_{\mathbf{X}}$, so that the orientation of time is the same throughout the interval.

The final postulate of CSM is that, in actual physical systems, the boundary condition is set at a single point in time.[12] This is not just true for the cosmos as a whole, I claim, but also for the subsystems within it, including time slices of systems. I shall argue for this postulate by consideration of theoretical elegance, in this section, and on empirical grounds, by showing that it has the right physical consequences, in the next section.

**CSM5** The boundary condition of a system $\mathbf{X}$ specifies the state of $\mathbf{X}$ at one instant of time, that is, $bc_{\mathbf{X}}$ is of the form $\mathbf{X}(t_1) = x_1$.

The first argument for CSM5 is that it allows CSM4 to be true, as will now be shown. The following notion of a connected partition will be useful here.

**5.6.1 Definition** A *connected partition* of $T_\mathbf{X}$ is a class of closed sub-intervals of $T_\mathbf{X}$ whose union is $T_\mathbf{X}$ and whose pairwise intersection is either empty or a singleton set.

Thus, for instance, if $T_\mathbf{X}$ is [0, 10], then one connected partition of $T_\mathbf{X}$ is {[0, 3], [3, 4], [4, 10]}.

Consider this connected partition of a stochastic system **X**. If the boundary condition is at a single point of time, then it constrains only one time slice of the partition (unless it happens to hit the point of intersection of two adjacent intervals, a case that we consider below). For example, let us suppose that the boundary condition is set at $t = 3.1$. The time slice [3, 4] then has a lagrangian and a boundary condition, which are sufficient to cause its actual history. Thus, the actual history of this time slice comes into being. Now, since the history in [3, 4] is actual, the states at $t = 3$ and $t = 4$ are actual as well, since they are sub-histories within this interval. The state at $t = 3$, however, is also part of the interval [0, 3], so the interval [0, 3] has one of its states constrained by the actual history in [3, 4]. This constraint on [0, 3] is a boundary condition for it, so it has a sufficient cause for its own actual history, which also comes into being. In a similar way, the state of [3, 4] at $t = 4$ provides a boundary condition for the time slice in [4, 10].

If the boundary condition falls on the intersection of two time slices, such as at $t = 3$, then it causes histories for both time slices [0, 3] and [3, 4]. The actual history in [3, 4] then provides a boundary condition for [4, 10] as before.

We see then that the time slice which has the boundary condition for the total system **X** provides a (single-time) boundary condition for each of its neighbours. These, in turn, by acquiring actual histories, provide boundary conditions for their adjacent time slices, and so on. Thus, each time slice receives its own single-time boundary condition, and hence CSM4 is satisfied. It is hard to conceive of any other kind of boundary condition that will yield this same result. For a *particular* connected partition of **X** we can no doubt cook up such a boundary condition, but this is uninteresting since any partition of **X** is arbitrary, and so if CSM4 is valid it should hold for *all* such partitions.

The second argument for CSM5 is that it (together with the other postulates) enables one to prove the Markov property, which (intuitively) ought to follow from CSM3 and CSM4 together. Moreover, closed systems are usually assumed to be Markov, since in that case the system cannot have an external 'memory.' Before we can show this, however, we need some preliminary results.

**5.6.2 Theorem** The law function is *point normalized*, that is, if $C$ is a subset of $H_{\mathbf{X}}$ containing all the histories $\boldsymbol{x}$ such that $\boldsymbol{x}(t') = x$, for fixed $t'$, $x$, then $L_{\mathbf{X}}(C) = 1$.

*Proof*: $L_{\mathbf{X}}(C) = Pr(C \mid \mathcal{L}_{\mathbf{X}} \;\&\; \mathbf{X}(t') = x)$, by Definition 5.2.1, since each member of $C$ satisfies the boundary condition $\mathbf{X}(t') = x$. But, since $C$ is the class of *all* histories in $H_{\mathbf{X}}$ satisfying this condition, $Pr(C \mid \mathcal{L}_{\mathbf{X}} \;\&\; \mathbf{X}(t') = x) = 1$.■

**5.6.3 Definition** Suppose $\{T_i\} = \{T_1, T_2, \ldots, T_n\}$ is a connected partition of $T_{\mathbf{X}}$. Then possible histories in $T_i$ will be denoted $\boldsymbol{a}_i$, $\boldsymbol{b}_i$, etc. and variables over these $\boldsymbol{x}_i$. The lagrangian for $T_i$ is denoted $\mathcal{L}_i$, and the corresponding law function is $L_i$.

The main result that is used to prove the Markov property is a product rule for time slices, analogous to Theorem 5.4.1, as follows.

**5.6.4 Theorem** Consider a connected partition $\{T_i\}$ of $T_{\mathbf{X}}$ that splits the history $\boldsymbol{a}$ into connected sub-histories $\{\boldsymbol{a}_i\}$. Then $L_{\mathbf{X}}(\boldsymbol{a}) = L_1(\boldsymbol{a}_1)L_2(\boldsymbol{a}_2) \ldots L_n(\boldsymbol{a}_n)$.

*Proof*: We only need consider the case $n = 2$, as then the other cases follow by iteration. Let $t_1$ be the single member of $T_1 \cap T_2$, so that $\boldsymbol{a}_1(t_1) = \boldsymbol{a}_2(t_1)$. The relation $L_{\mathbf{X}}(\boldsymbol{a}) = L_1(\boldsymbol{a}_1)L_2(\boldsymbol{a}_2)$ is true just in case the relation $P_{\mathbf{X}}(\boldsymbol{a}) = P_1(\boldsymbol{a}_1)P_2(\boldsymbol{a}_2)$ is true when a satisfies $bc_{\mathbf{X}}$, $\boldsymbol{a}_1$ satisfies $bc_1$, and $\boldsymbol{a}_2$ satisfies $bc_2$. Let us therefore assume that $\boldsymbol{a}$, $\boldsymbol{a}_1$, and $\boldsymbol{a}_2$ all satisfy their respective boundary conditions. We further assume wlog that $bc_{\mathbf{X}}$ falls within $T_1$, so that $bc_{\mathbf{X}}$ constrains $\mathbf{X}_1$ only, and thus is equivalent to $bc_1$. We then have

$$
\begin{aligned}
P_{\mathbf{X}}(\boldsymbol{a}) &= P_{\mathbf{X}}(\boldsymbol{a}_1 \;\&\; \boldsymbol{a}_2) \\
&= Pr(\boldsymbol{a}_1 \;\&\; \boldsymbol{a}_2 \mid \mathcal{L}_{\mathbf{X}} \;\&\; bc_{\mathbf{X}}) \\
&= Pr(\boldsymbol{a}_1 \;\&\; \boldsymbol{a}_2 \mid \mathcal{L}_1 \;\&\; \mathcal{L}_2 \;\&\; bc_1) \\
&= Pr(\boldsymbol{a}_1 \mid \mathcal{L}_1 \;\&\; \mathcal{L}_2 \;\&\; bc_1)Pr(\boldsymbol{a}_2 \mid \boldsymbol{a}_1 \;\&\; \mathcal{L}_1 \;\&\; \mathcal{L}_2 \;\&\; bc_1) \\
&= Pr(\boldsymbol{a}_1 \mid \mathcal{L}_1 \;\&\; bc_1)Pr(\boldsymbol{a}_2 \mid \boldsymbol{a}_1(t_1) \;\&\; \mathcal{L}_2) \\
&= P_1(\boldsymbol{a}_1)P_2(\boldsymbol{a}_2).
\end{aligned}
$$

(Note that this proof is exactly parallel to the proof of Theorem 5.4.3, and also relies on CSM3. It is necessary to assume, for instance, that $\boldsymbol{a} = \boldsymbol{a}_1 \;\&\; \boldsymbol{a}_2$.) Since the assumption that $\boldsymbol{a}$, $\boldsymbol{a}_1$, and $\boldsymbol{a}_2$ all satisfy their respective boundary conditions leads to the conclusion that $P_{\mathbf{X}}(\boldsymbol{a}) = P_1(\boldsymbol{a}_1)P_2(\boldsymbol{a}_2)$, it follows that $L_{\mathbf{X}}(\boldsymbol{a}) = L_1(\boldsymbol{a}_1)L_2(\boldsymbol{a}_2)$.■

The intuitive idea of the law function is a measure of dynamical facility, of how easily the system follows the trajectory in question. Now, the dynamical facility of a sub-history such as $\boldsymbol{a}_2$ should depend only upon the lagrangian $\mathcal{L}_2$, so it should be the case that $L_{\mathbf{X}}(\boldsymbol{a}_2) = L_2(\boldsymbol{a}_2)$. We shall now prove this.

**5.6.5 Theorem** $L_{\mathbf{X}}(\boldsymbol{a}_i) = L_i(\boldsymbol{a}_i)$.

*Proof*: Let us consider the connected partition $\{T_l, T_i, T_u\}$ of $T_{\mathbf{X}}$, where '$l$' and '$u$' should be read 'lower' and 'upper.' We then have that

$$L_{\mathbf{X}}(\boldsymbol{a}_i) = \sum_{\boldsymbol{x}_l, \boldsymbol{x}_u} L_{\mathbf{X}}(\boldsymbol{x}_l \;\&\; \boldsymbol{a}_i \;\&\; \boldsymbol{x}_u)$$

where the variables $\boldsymbol{x}_l$ and $\boldsymbol{x}_u$ range over all histories in $T_l$, $T_u$ respectively that connect to $\boldsymbol{a}_i$. Using Theorem 5.6.4 it follows that

$$\begin{aligned} L_{\mathbf{X}}(\boldsymbol{a}_i) &= \sum_{\boldsymbol{x}_l, \boldsymbol{x}_u} L_l(\boldsymbol{x}_l)L_i(\boldsymbol{a}_i)L_u(\boldsymbol{x}_u) \\ &= L_i(\boldsymbol{a}_i) \sum_{\boldsymbol{x}_l} L_l(\boldsymbol{x}_l) \sum_{\boldsymbol{x}_u} L_u(\boldsymbol{x}_u) \\ &= L_i(\boldsymbol{a}_i). \end{aligned}$$

The last line follows using point normalization (Theorem 5.6.2), since the variables $\boldsymbol{x}_l$ and $\boldsymbol{x}_u$ are each constrained at a single instant of time.■

**5.6.6 Corollaries** (i) If $bc_{\mathbf{X}}$ is set in $T_i$, and consistent with $\boldsymbol{a}_i$, then $P_{\mathbf{X}}(\boldsymbol{a}_i) = L_{\mathbf{X}}(\boldsymbol{a}_i)$.
(ii) If $bc_{\mathbf{X}}$ is set in $T_l$, then $P_{\mathbf{X}}(\boldsymbol{a}_i) = L_i(\boldsymbol{a}_i) \sum_{\boldsymbol{x}_l} L_l(\boldsymbol{x}_l)$.

We are finally in a position to prove the Markov condition.

**5.6.7 Theorem** Consider sub-histories $\boldsymbol{a}_1$, $\boldsymbol{a}_2$, $\boldsymbol{a}_3$ in time intervals $T_1$, $T_2$, $T_3$ respectively, where $T_1 < T_2 < T_3$. (These intervals may or may not form a connected partition of $T_{\mathbf{X}}$.) Then $P_{\mathbf{X}}(\boldsymbol{a}_1 \;\&\; \boldsymbol{a}_3 \mid \boldsymbol{a}_2) = P_{\mathbf{X}}(\boldsymbol{a}_1 \mid \boldsymbol{a}_2)P_{\mathbf{X}}(\boldsymbol{a}_3 \mid \boldsymbol{a}_2)$.

This is easily proved from the following lemma.

**5.6.8 Lemma** Consider the connected partition $\{T_l, T_2, T_u\}$ of $T_{\mathbf{X}}$, so that $T_1 \subseteq T_l$ and $T_3 \subseteq T_u$. Then, for all sub-histories $\boldsymbol{x}_l$, $\boldsymbol{x}_u$ in $T_l$, $T_u$, $P_{\mathbf{X}}(\boldsymbol{x}_l \,\&\, \boldsymbol{x}_u \mid \boldsymbol{a}_2) = P_{\mathbf{X}}(\boldsymbol{x}_l \mid \boldsymbol{a}_2)P_{\mathbf{X}}(\boldsymbol{x}_u \mid \boldsymbol{a}_2)$.

*Proof*: From Axiom 4 we have

$$P_{\mathbf{X}}(\boldsymbol{x}_l \,\&\, \boldsymbol{x}_u \mid \boldsymbol{a}_2) = \frac{P_{\mathbf{X}}(\boldsymbol{x}_l \,\&\, \boldsymbol{a}_2 \,\&\, \boldsymbol{x}_u)}{P_{\mathbf{X}}(\boldsymbol{a}_2)}.$$

Case (1): Suppose that $bc_{\mathbf{X}}$ is set somewhere in $T_l$. Then, for every $\boldsymbol{x}_l$ that satisfies $bc_{\mathbf{X}}$,

$$\begin{aligned} P_{\mathbf{X}}(\boldsymbol{x}_l \,\&\, \boldsymbol{x}_u | \boldsymbol{a}_2) &= \frac{L_{\mathbf{X}}(\boldsymbol{x}_l \,\&\, \boldsymbol{a}_2 \,\&\, \boldsymbol{x}_u)}{\sum\limits_{\mathbf{y}_l} P_{\mathbf{X}}(\boldsymbol{y}_l \,\&\, \boldsymbol{a}_2)} \\ &= \frac{L_l(\boldsymbol{x}_l)L_2(\boldsymbol{a}_2)L_u(\boldsymbol{x}_u)}{\sum\limits_{\mathbf{y}_l} L_l(\boldsymbol{y}_l)L_2(\boldsymbol{a}_2)} \\ &= \frac{L_l(\boldsymbol{x}_l)L_u(\boldsymbol{x}_u)}{\sum\limits_{\mathbf{y}_l} L_l(\boldsymbol{y}_l)}. \end{aligned}$$

Note that $\boldsymbol{y}_l$ is constrained to agree with $\boldsymbol{a}_2$ at their point of intersection, and also to satisfy $bc_{\mathbf{X}}$, so the denominator is less than one. By similar reasoning we find that

$$P_{\mathbf{X}}(\boldsymbol{x}_l | \boldsymbol{a}_2) = \frac{L_l(\boldsymbol{x}_l)}{\sum\limits_{\boldsymbol{y}_l} L_l(\boldsymbol{y}_l)}$$

$$P_{\mathbf{X}}(\boldsymbol{x}_u | \boldsymbol{a}_2) = L_u(\boldsymbol{x}_u),$$

and hence $P_{\mathbf{X}}(\boldsymbol{x}_l \,\&\, \boldsymbol{x}_u \mid \boldsymbol{a}_2) = P_{\mathbf{X}}(\boldsymbol{x}_l \mid \boldsymbol{a}_2) P_{\mathbf{X}}(\boldsymbol{x}_u \mid \boldsymbol{a}_2)$.

Case (2): $bc_{\mathbf{X}}$ is set somewhere in $T_2$. It is trivial to show that $P_{\mathbf{X}}(\boldsymbol{x}_l \,\&\, \boldsymbol{x}_u \mid \boldsymbol{a}_2) = P_{\mathbf{X}}(\boldsymbol{x}_l \mid \boldsymbol{a}_2) P_{\mathbf{X}}(\boldsymbol{x}_u \mid \boldsymbol{a}_2)$.

Case (3): $bc_{\mathbf{X}}$ is set somewhere in $T_u$. From case (1), $P_{\mathbf{X}}(\boldsymbol{x}_l \,\&\, \boldsymbol{x}_u \mid \boldsymbol{a}_2) = P_{\mathbf{X}}(\boldsymbol{x}_l \mid \boldsymbol{a}_2) P_{\mathbf{X}}(\boldsymbol{x}_u \mid \boldsymbol{a}_2)$ follows by symmetry.

Thus, we infer that $P_{\mathbf{X}}(\boldsymbol{x}_l \,\&\, \boldsymbol{x}_u \mid \boldsymbol{a}_2) = P_{\mathbf{X}}(\boldsymbol{x}_l \mid \boldsymbol{a}_2) P_{\mathbf{X}}(\boldsymbol{x}_u \mid \boldsymbol{a}_2)$, as required.■

*Proof* of Theorem 5.6.7:

$$
\begin{aligned}
P_{\mathbf{X}}(\boldsymbol{a}_1 \,\&\, \boldsymbol{a}_3 | \boldsymbol{a}_2) &= \sum_{\boldsymbol{x}_l \supseteq \boldsymbol{a}_1} \sum_{\boldsymbol{x}_u \supseteq \boldsymbol{a}_3} P_{\mathbf{X}}(\boldsymbol{x}_l \,\&\, \boldsymbol{x}_u | \boldsymbol{a}_2) \\
&= \sum_{\boldsymbol{x}_l \supseteq \boldsymbol{a}_1} \sum_{\boldsymbol{x}_u \supseteq \boldsymbol{a}_3} P_{\mathbf{X}}(\boldsymbol{x}_l | \boldsymbol{a}_2) P_{\mathbf{X}}(\boldsymbol{x}_u | \boldsymbol{a}_2) \\
&= P_{\mathbf{X}}(\boldsymbol{a}_1 | \boldsymbol{a}_2) P_{\mathbf{X}}(\boldsymbol{a}_3 | \boldsymbol{a}_2).
\end{aligned}
$$

This is the required result.■

The chance of an event does not vary with time. Yet, since chance is relativized to a system, and time slices are systems, we can easily define a time-dependent chance, which I call *up-to-date* (utd) chance. The basic idea is that the chance at time $t$ of some later event $E$ is the chance of $E$ within the time slice system in $[t, \tau]$. Note that this depends on the actual history before $t$. It is not predictable in advance of $t$ what the chance of $E$ at $t$ will be.

**5.6.9 Definition** Consider a sub-history $\boldsymbol{x}_i$ in $T_i$, and a time $t < T_i$, so that $T_i \subset [t, \tau]$. Then the chance at time $t$ of $\boldsymbol{x}_i$, written $P_t(\boldsymbol{x}_i)$, is the chance of $\boldsymbol{x}_i$ within the time slice $[t, \tau]$.

As one might expect, utd chances are related to conditional chances, as the following results show. The interval $T_r$ is $[t, t_i]$, where $t_i$ is the lower bound of $T_i$.

**5.6.10 Lemma** $P_t(\boldsymbol{x}_i) = L_i(\boldsymbol{x}_i) \sum_r L_r(\boldsymbol{x}_r)$, where the $\boldsymbol{x}_r$ are all consistent with $\boldsymbol{x}_i$ and $\mathbf{X}(t)$.

*Proof*:

$$\begin{aligned} P_t(\boldsymbol{x}_i) &= \sum_r L_t(\boldsymbol{x}_r \,\&\, \boldsymbol{x}_i) \\ &= \sum_r L_r(\boldsymbol{x}_r) L_i(\boldsymbol{x}_i) \\ &= L_i(\boldsymbol{x}_i) \sum_r L_r(\boldsymbol{x}_r). \end{aligned}$$

**5.6.11 Theorem** If $\mathbf{X}[0, \mathrm{t}]$ specifies the actual history of $\mathbf{X}$ in $[0, t]$, then $P_t(\boldsymbol{x}_i) = P_{\mathbf{X}}(\boldsymbol{x}_i \mid \mathbf{X}[0, \mathrm{t}])$.

*Proof*: Let the actual history in $[0, t]$ be $\boldsymbol{a}_0$. Then

$$\begin{aligned} P_{\mathbf{X}}(\boldsymbol{x}_i | \boldsymbol{a}_0) &= \frac{P_{\mathbf{X}}(\boldsymbol{a}_0 \,\&\, \boldsymbol{x}_i)}{P_{\mathbf{X}}(\boldsymbol{a}_0)} \\ &= \frac{\sum_r P_{\mathbf{X}}(\boldsymbol{a}_0 \,\&\, \boldsymbol{x}_r \,\&\, \boldsymbol{x}_i)}{P_{\mathbf{X}}(\boldsymbol{a}_0)} \\ &= \frac{\sum_r L_0(\boldsymbol{a}_0) L_r(\boldsymbol{x}_r) L_i(\boldsymbol{x}_i)}{L_0(\boldsymbol{a}_0)} \\ &= L_i(\boldsymbol{x}_i) \sum_r L_r(\boldsymbol{x}_r). \end{aligned}$$

Using Lemma 5.6.10, it then follows that $P_t(\boldsymbol{x}_i) = P_{\mathbf{X}}(\boldsymbol{x}_i \mid \mathbf{X}[0, \mathrm{t}])$, as required.■

**5.6.12 Corollary** $P_t(\boldsymbol{x}_i) = P_{\mathbf{X}}(\boldsymbol{x}_i \mid \mathbf{X}(t))$.

*Proof*: Immediate from theorems 5.6.11 and 5.6.7.

Note that, since $\boldsymbol{x}_i$ may be a state, a utd chance may be a forward-transition probability.

## 5.7 THE ARROW OF TIME

In the previous section we saw that postulate CSM5, which says that the physical boundary condition of every system is a constraint at a single point of time, has two theoretical virtues: it allows CSM4 to be true, and together with CSM4 entails the Markov relation. In this section I shall argue that a single-instant boundary condition also provides a suitable arrow of time.

What are the phenomena that constitute the 'arrow(s) of time'? The three such phenomena considered in this chapter are, as mentioned in §5.1:

1 Traces of an event (such as a crater, with respect to a meteor impact) may only succeed, not precede that event.
2 Forward, but not backward, transition probabilities are lawlike and time-independent.
3 Entropy tends to increase.

Other such phenomena are sometimes discussed as arrows of time, such as the expansion of the cosmos and the outward propagation of radiation, but these three are sufficient to demonstrate how CSM can be used to investigate the arrow of time.

The puzzle created by these phenomena lies in the fact that the laws of physics are time-symmetric. If, therefore, one explains phenomena only by these laws, then one is in the difficult position of explaining an asymmetric effect by appeal to a symmetric cause. Of course, if the cosmos is stochastic, then the actual history may turn out to be time-asymmetric just by chance. However, the fact that there are several distinct arrows of time, which are all consistent with one another, strongly suggests that the actual history has some cause that is not time-symmetric.[13]

In my view there is such a time-asymmetric cause, which is the boundary condition. The boundary condition is not time-asymmetric by virtue of the state that it fixes, but rather from being a constraint on the cosmos *at one instant of time*. Any such boundary condition, regardless of the state involved, will cause phenomena (1) and (2). Moreover, if the fixed state is also one of low entropy, then (3) is brought about as well.

### *5.7.1 Time's Arrow and Causation*

In the previous section it was shown how, if the boundary condition constrains the actual history at just one instant, one time slice may provide a boundary condition for its neighbour. In the special case where the boundary condition is set at either extreme of $T_{\mathbf{X}}$, this means that the time slices of **X** are well ordered by the cause–effect relation, as will now be shown.

Suppose $T_{\mathbf{X}} = [0, 10]$ and $bc_{\mathbf{X}}$ is set at $t = 10$ (it would just as well if set at $t = 0$). Also, let us consider the arbitrary connected partition {[0, 1], [1, 3], [3, 8], [8, 10]} of $T_{\mathbf{X}}$. Since the time slice in [8, 10] has a boundary condition, it has a sufficient cause for its actual history, so an actual history for [8, 10] comes into being. This actual history fixes the state of **X** at $t = 8$, which is part of the interval [3, 8], so then this time slice acquires a boundary condition. Having a boundary condition, its actual history is also caused to exist, which in turn sets a boundary condition for [1, 3] and so on. The actual history in [8, 10] is brought about by the boundary condition, and this history causes the history in [3, 8], which causes the history in [1, 3], which causes the actual history in [0, 1].

There is nothing special about the partition chosen; the same causal ordering would hold for any such partition. Also, I considered a boundary condition at $t = 10$ rather than $t = 0$, making causation run 'backwards,' to show that the causal order does not depend on the <-ordering of the real numbers. Note that, on this approach, the coincidence of temporal order with causal order is not analytic but merely a physical fact. If the boundary condition is set at some time that is not an extremum of $T_{\mathbf{X}}$, for example, then the cause–effect relation does not provide even a linear ordering of the time slices. Moreover, if the boundary condition constrained the actual history at two or more different times, then presumably an even stranger situation would obtain.

In the above example there is no mention of any causal ordering within a single time slice, such as [3, 8]. Since CSM4 allows *any* partition into time slices, however, we can consider the time slice in [3, 8] to be composed of two subsystems, one in [3, 5] and the other in [5, 8] perhaps. From this perspective we see that the actual history in [5, 8] causes the history in [3, 5], and so a little more of the causal order is revealed. Since this causal structure will appear for any connected partition of $T_{\mathbf{X}}$, no matter how fine, we see that there is in fact a continuous causal flow of events (in a closed system) rather than a discrete causal chain.

The above talk of a causal order of becoming and a continuous causal flow may be taken to suggest that I have abandoned my earlier commitment to a 'B-series' view of time. This is not the case. The recognition of a causal structure in spacetime does not require one to believe in a moving present. I do hope, however, that this recognition may weaken the common intuition that the B-series view leaves out something important about time that distinguishes it from a mere physical dimension. Causation does indeed relate to time and space quite differently, and there really is something that may be called a flow of potentiality into actuality. Perhaps the so-called ‘flow’ of time is really just the causal ordering.

In this chapter I am concerned with time rather than with space, so the three spatial dimensions have been ignored altogether. If they were included, then we would have to deal with spacetime slices rather than time slices, and each spacetime slice would be a cause of the other slices that are in its future light cone. Thus, although I am currently ignoring space, it seems that my account is capable of being extended to cover a four-dimensional cosmos.

This idea of a causal arrow of time that arises from a single-instant boundary condition is not often taken seriously, even though it fits very well with our pre-theoretical intuitions about time. Why is this? I think there are two main reasons. First, to explain the arrows of time by means of a physical constraint on the whole cosmos seems like resorting to a *deus ex machina*. One would prefer to explain all phenomena by means of the intrinsic properties of the cosmos itself, with no interference from ‘outside.’ Second, the very simplicity of the hypothesis, and its conformity with common sense, can be seen as weaknesses. It may be considered merely a ‘folk’ explanation of the phenomenon, involving pre-scientific notions. A scientific explanation would be more technical and sophisticated, and would make no fundamental use of the antiquated idea of causation.

With regard to the first objection, I agree that a ‘self-contained’ explanation of the arrow of time would be preferable, in accordance with Ockham’s Razor. However, the very nature of dynamical systems strongly suggests that they do not contain within themselves the entire cause of their own motions. The intrinsic nature of a system includes what we may call laws of development, governing *transitions* from one state to another, but it does not (we generally think) impose any lawlike constraint on the initial state, or any other state. The appeal to an external constraint is therefore unavoidable.

As for the second objection, we have seen that the old idea of cause and effect is needed to give a theory of physical chance. Attempts to eliminate the notion of causation (or reduce it to something less mysterious) have failed, and there is little prospect for success in the future. If there really is a causal relation, then it must be deeply involved in the arrow of time. Since the causal relation itself cannot be reduced to anything more basic, it is unlikely that its role in this phenomenon can be expressed in other, more modern, terms.

### *5.7.2 Temporal Symmetries of the Lagrangian*

Physical laws are generally thought to possess a high degree of temporal symmetry. In CSM the physical laws are represented by $\mathscr{L}$, the lagrangian, so we might expect $\mathscr{L}$ to display some temporal symmetry as well. So far we have only postulated one such kind of symmetry for $\mathscr{L}$, but it is very easy to impose two further kinds. There are two reasons to postulate that $\mathscr{L}$ has these temporal symmetries: First, there is evidence that real systems do have the symmetries in question. Second, the postulation of these symmetries highlights the fact that the temporal asymmetries of $P$ are due entirely to CSM5, that the boundary condition constrains the state at a single time, and not to any temporal asymmetry of $\mathscr{L}$.

The two kinds of time symmetry in $\mathscr{L}$ are expressed in terms of the law function $L$. They are usually called *time-translation invariance* and *time-reversal invariance*. The basic idea of time-translation invariance is that, for a given system **X**, $\mathscr{L}_{\mathbf{X}}$ does not vary with time. Thus, if $\mathscr{L}_1$ and $\mathscr{L}_2$ are lagrangians of equal-duration time slices of the same system, then $\mathscr{L}_1 = \mathscr{L}_2$. In terms of the law function we have the following definition.

**5.7.2.1 Definition** The lagrangian $\mathscr{L}_{\mathbf{X}}$ is time-translation invariant (tti) just in case for any two equal-duration time slices of **X**, such as $\mathbf{X}_1$ and $\mathbf{X}_2$, and any possible history $\boldsymbol{a}$ of $\mathbf{X}_1$, $L_1(\boldsymbol{a}) = L_2(\boldsymbol{a})$.

The basic idea of time-reversal invariance is that the lagrangian $\mathscr{L}$ does not distinguish between the two directions of time. To make this notion more precise we require the concept of a time-reflection, or a time-reverse, of a possible history. If we imagine that a possible history is some sort of geometrical object, then it can be reflected in a 'simultaneity plane.' The image $\boldsymbol{a}^r$ under reflection of a history $\boldsymbol{a}$ will be different from $\boldsymbol{a}$, and yet the two will be congruent. Intuitively, $\boldsymbol{a}^r$ is just $\boldsymbol{a}$ going backwards.

**5.7.2.2 Definition** The lagrangian $\mathscr{L}_{\mathbf{X}}$ is time-reversal invariant (tri) just in case, for any possible history $\boldsymbol{a}$ of **X**, $L_{\mathbf{X}}(\boldsymbol{a}) = L_{\mathbf{X}}(\boldsymbol{a}^r)$.

The property of tti entails that forward transition chances are independent of time, that is,

$$P_{\mathbf{X}}(\mathbf{X}(t_1 + d) = b \mid \mathbf{X}(t_1) = a) = P_{\mathbf{X}}(\mathbf{X}(t_2 + d) = b \mid \mathbf{X}(t_2) = a).$$

The property of tri entails such relations as

$$P_{\mathbf{X}}(\mathbf{X}(t_2) = b \mid \mathbf{X}(t_1) = a) = P_{\mathbf{X}}(\mathbf{X}(t_2) = a^r \mid \mathbf{X}(t_1) = b^r).$$

### 5.7.3 *Time's Arrow and Chance*

We have supposed that $\mathcal{L}$ has every conceivable kind of time symmetry, and these symmetries show up as time symmetries in $P$; thus, $P$ is also highly symmetric with respect to time. Let us now fix our time coordinates so that the boundary condition is set at $t = 0$, and only consider positive times, so that $T_{\mathbf{X}}$ is the interval $[0, \tau]$.[14] There are then some important temporal asymmetries in $P$, of which we shall now examine the following four.

1 The forward transition chances are independent of time, whereas the backward ones are not.[15]
2 Interference with a system in a time interval $T_i$ alters the chance distributions for time slices after $T_i$, but not before $T_i$.
3 Diffusion occurs toward the future, but not toward the past.
4 Reichenbach's Common Cause Principle holds; that is, there exist conjunctive forks open to the future, but not open to the past.

Perhaps a better way of saying (1) is that the forward transition chances depend upon the lagrangian alone, whereas the backward transition chances depend upon the boundary condition as well as the lagrangian. This is why the backward transition chances are time-dependent; they vary with the proximity of the boundary condition, and this proximity depends on the time. It also accords with the common view that the forward transition chances are lawlike and 'physical,' whereas the backward ones are suspect. We can infer, from Lem. 5.6.10 and Cor. 5.6.12, that

$$P_{\mathbf{X}}(\boldsymbol{x}_i \mid \mathbf{X}(t) = a) = L_i(\boldsymbol{x}_i) \sum_r L_r(\boldsymbol{x}_r),$$

where $\boldsymbol{x}_i$ is a sub-history in $[t_i, \tau]$ and the $\boldsymbol{x}_r$ are all the sub-histories in $[t, t_i]$ such that $\boldsymbol{x}_r(t) = a$ and $\boldsymbol{x}_r(t_i) = \boldsymbol{x}_i(t_i)$.

Now, if we contract the sub-history $\boldsymbol{x}_i$ to a single state $b$, at time $t + d$ say, then $L_i(\boldsymbol{x}_i) = 1$. We thus infer that

$$P_{\mathbf{X}}(\mathbf{X}(t + d) = b \mid \mathbf{X}(t) = a) = \sum_r L_r(\boldsymbol{x}_r),$$

where the histories $\boldsymbol{x}_r$ in $[t, t + d]$ are all such that $\boldsymbol{x}_r(t) = a$ and $\boldsymbol{x}_r(t + d) = b$. Intuitively, the forward transition chance is just the law value of the class of histories that contain the transition in question from $a$ to $b$.

Since we have assumed that the law function is time-translation invariant, it immediately follows that the forward transition chances are independent of time.

**5.7.3.1 Lemma** Let $\boldsymbol{x}_i$ be a sub-history in $[t_i, \tau]$, where $\boldsymbol{x}_i(t_i) = a$. Then $P_{\mathbf{X}}(\boldsymbol{x}_i \mid \mathbf{X}(t_i) = a) = L_i(\boldsymbol{x}_i)$.

*Proof*: In the above relation, $P_{\mathbf{X}}(\boldsymbol{x}_i \mid \mathbf{X}(t) = a) = L_i(\boldsymbol{x}_i) \sum_{x_r} L_r(\boldsymbol{x}_r)$, let $t = t_i$, so that the sub-histories $\boldsymbol{x}_r$ all shrink to the state $a$ at time $t_i$. The law value of a state is one, so that the relation becomes $P_{\mathbf{X}}(\boldsymbol{x}_i \mid \mathbf{X}(t_i) = a) = L_i(\boldsymbol{x}_i)$, as required.■

**5.7.3.2 Theorem** The forward transition chances are independent of time; i.e., for any times $t_1, t_2 \in T_{\mathbf{X}}$, $P_{\mathbf{X}}(\mathbf{X}(t_1 + d) = b \mid \mathbf{X}(t_1) = a) = P_{\mathbf{X}}(\mathbf{X}(t_2 + d) = b \mid \mathbf{X}(t_2) = a)$.

*Proof*: Consider the two sub-intervals of $T_{\mathbf{X}}$, $T_1 = [t_1, t_1 + d]$ and $T_2 = [t_2, t_2 + d]$, which are of equal duration, and a sub-history $\boldsymbol{x}$ in $[t_1, t_1 + d]$ where $\boldsymbol{x}(t_1) = a$ and $\boldsymbol{x}(t_1 + d) = b$. By Lem. 5.7.3.1 we have that $P_{\mathbf{X}}(\boldsymbol{x} \mid \mathbf{X}(t_1) = a) = L_1(\boldsymbol{x})$. Similarly, if $\boldsymbol{x}'$ is the time translation of $\boldsymbol{x}$ into $T_2$, we have that $P_{\mathbf{X}}(\boldsymbol{x}' \mid \mathbf{X}(t_2) = a) = L_2(\boldsymbol{x}')$. Thus, since $P_{\mathbf{X}}(\mathbf{X}(t_1 + d) = b \mid \mathbf{X}(t_1) = a)$ and $P_{\mathbf{X}}(\mathbf{X}(t_2 + d) = b \mid \mathbf{X}(t_2) = a)$ are just sums of terms like $P_{\mathbf{X}}(\boldsymbol{x} \mid \mathbf{X}(t_1) = a)$ and $P_{\mathbf{X}}(\boldsymbol{x}' \mid \mathbf{X}(t_2) = a)$, and by tti $L_1(\boldsymbol{x}) = L_2(\boldsymbol{x}')$, it follows that $P_{\mathbf{X}}(\mathbf{X}(t_1 + d) = b \mid \mathbf{X}(t_1) = a) = P_{\mathbf{X}}(\mathbf{X}(t_2 + d) = b \mid \mathbf{X}(t_2) = a)$.■

**5.7.3.3 Theorem** The backward transition chances depend on the boundary condition, and are not independent of time.

*Proof*: Let us consider the backward conditional chance $P_{\mathbf{X}}(\boldsymbol{a}_1 \mid \boldsymbol{a}_2)$, where $\boldsymbol{a}_1$ is in $[t, t + d]$ and $\boldsymbol{a}_2$ is in $[t + d, t + d + d']$. We have that

$$P_{\mathbf{X}}(\boldsymbol{a}_1 | \boldsymbol{a}_2) = \frac{P_{\mathbf{X}}(\boldsymbol{a}_1 \ \&\ \boldsymbol{a}_2)}{P_{\mathbf{X}}(\boldsymbol{a}_2)}$$

$$= \frac{\sum_{\boldsymbol{x}_0} P_{\mathbf{X}}(\boldsymbol{x}_0 \ \&\ \boldsymbol{a}_1 \ \&\ \boldsymbol{a}_2)}{\sum_{\boldsymbol{x}_0, \boldsymbol{x}_1} P_{\mathbf{X}}(\boldsymbol{x}_0 \ \&\ \boldsymbol{x}_1 \ \&\ \boldsymbol{a}_2)}$$

$$= \frac{\sum_{x_0} L_0(x_0)L_1(a_1)L_2(a_2)}{\sum_{x_0,x_1} L_0(x_0)L_1(x_1)L_2(a_2)}$$

$$= \frac{L_1(a_1)\sum_{x_0} L_0(x_0)}{\sum_{x_0,x_1} L_0(x_0)L_1(x_1)}.$$

If the temporal order of $a_1$ and $a_2$ were reversed, then $P_X(a_1 \mid a_2)$ would just be $L_1(a_1)$. Thus, we see that the backward transition probability has two additional terms, one in the numerator and one in the denominator. Both of these depend on the boundary condition, and thus on the time.■

Now let us prove property (2). Consider the chance function for a system **X** that suffers outside interference within some time interval $T_i$. The standard example of this is where **X** is a patch of sand, and the interference consists of a foot stepping on it. Observations indicate that after the footstep there is a foot-shaped impression in the sand that did not exist before. The difficulty is to explain the time asymmetry of this phenomenon.

We represent the fact that the system **X** is disturbed during $T_i$ in the usual way, by modifying $\mathcal{L}_i$ and thus $L_i$.

**5.7.3.4 Theorem** (a) The chance function *before* $T_i$ is independent of the interaction in $T_i$.
(b) The chance function *after* $T_i$ is dependent on the interaction in $T_i$.

*Proof*:
(a) Consider a history $a_1$ in $T_1 < T_i$. Let $T_0$ be the closed interval from 0 to $T_1$, and $T_2$ be the closed interval from $T_1$ to $T_i$, as in figure 5.1. Then

$$P_X(a_1) = \sum_{\substack{x_0,x_2,\\ x_i,x_3}} P_X(x_0 \,\&\, a_1 \,\&\, x_2 \,\&\, x_i \,\&\, x_3)$$

$$= \sum_{\substack{x_0, x_2, \\ x_i, x_3}} L_0(\boldsymbol{x}_0) L_1(\boldsymbol{a}_1) L_2(\boldsymbol{x}_2) L_i(\boldsymbol{x}_i) L_3(\boldsymbol{x}_3)$$

$$= L_1(\boldsymbol{a}_1) \sum_{x_0} L_0(\boldsymbol{x}_0) \sum_{x_2} L_2(\boldsymbol{x}_2) \sum_{x_i} L_i(\boldsymbol{x}_i) \sum_{x_3} L_3(\boldsymbol{x}_3)$$

Now, the variable $\boldsymbol{x}_0$ is constrained by the boundary condition, and also constrained to fit with $\boldsymbol{a}_1$ at the intersection of $T_0$ with $T_1$. The variable $\boldsymbol{x}_2$ is constrained at the lower end only, to fit with $\boldsymbol{a}_1$ at $T_1 \cap T_2$. In a similar way, $\boldsymbol{x}_i$ and $\boldsymbol{x}_3$ are constrained only at the lower end. Thus, by Theorem 5.6.2, $\Sigma_{\mathbf{x}_3} L_3(\mathbf{x}_3) = 1$, and may be cancelled from the summation. The same applies to all the other sums whose variables have only one constraint. This leaves us with

$$P_{\mathbf{X}}(\boldsymbol{a}_1) = L_1(\boldsymbol{a}_1) \sum_{x_0} L_0(\boldsymbol{x}_0),$$

which is clearly independent of $L_i$.

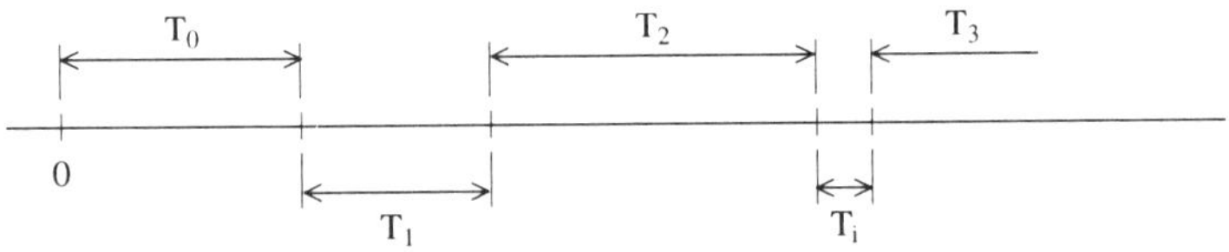


Figure 5.1

(b) Consider a history $\boldsymbol{a}_2$ in $T_2 > T_i$. This time let $T_0$ be the entire closed interval from 0 to $T_i$, and let $T_1$ be the closed interval from $T_i$ to $T_2$, as in figure 5.2. Then

$$P_{\mathbf{X}}(\boldsymbol{a}_2) = \sum_{\substack{x_0, x_i, \\ x_1, x_3}} P_{\mathbf{X}}(\boldsymbol{x}_0 \,\&\, \boldsymbol{x}_i \,\&\, \boldsymbol{x}_1 \,\&\, \boldsymbol{a}_2 \,\&\, \boldsymbol{x}_3)$$

$$= \sum_{\substack{x_0, x_i, \\ x_1, x_3}} L_0(\boldsymbol{x}_0) L_i(\boldsymbol{x}_i) L_1(\boldsymbol{x}_1) L_2(\boldsymbol{a}_2) L_3(\boldsymbol{x}_3)$$

$$= L_2(\boldsymbol{a}_2) \sum_{x_0} L_0(\boldsymbol{x}_0) \sum_{x_i} L_i(\boldsymbol{x}_i) \sum_{x_1} L_1(\boldsymbol{x}_1) \sum_{x_3} L_3(\boldsymbol{x}_3)$$

$$= L_2(\boldsymbol{a}_2) \sum_{x_0} L_0(\boldsymbol{x}_0) \sum_{x_i} L_i(\boldsymbol{x}_i) \sum_{x_1} L_1(\boldsymbol{x}_1) \cdot$$

Note that $\boldsymbol{x}_1$ is constrained by $\boldsymbol{x}_i$ and $\boldsymbol{a}_2$, and thus $\sum_{x_1} L_1(\boldsymbol{x}_1)$ is not equal to one, so that it cannot be cancelled. The same is then true of $\boldsymbol{x}_i$, so that the total expression depends on $L_i$.■

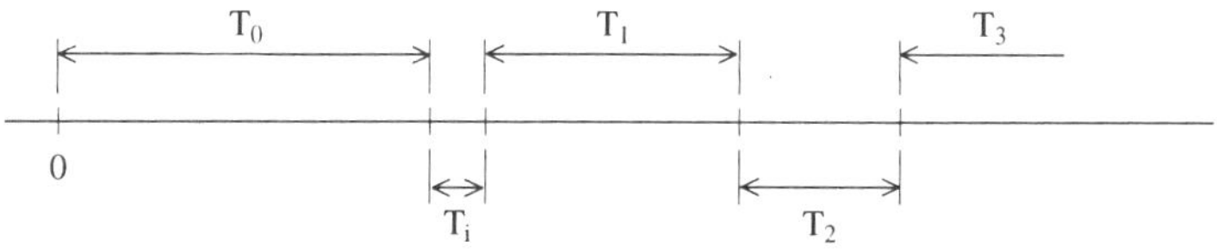


Figure 5.2

Now let us prove property (3), that diffusion occurs in the forward direction of time. To examine this issue we will need a model of a process in which diffusion can occur. The Ehrenfest urn model, though rather simple, is sufficient for our modest purpose.

The Ehrenfest urn model is a chamber with two compartments, separated by a slightly permeable membrane. The chamber contains $N$ particles, each of which is in one of the compartments. The particles are free to move within each compartment, and also have a small chance, at any time, of passing through the membrane into the other compartment. To keep things simple we will not consider the exact position and momentum of each particle, but only the compartment it occupies. Thus, a state of the system specifies the compartment ($A$ or $B$) of each of the $N$ particles. Also we will only consider the state of the system at integer times $t = 0, 1, 2, ...$, etc., so that if a particle penetrates the membrane twice in [2, 3], for example, this motion will not show up in the history. A history of the urn specifies the compartment occupied by each particle at all the times $\{0, 1, 2, ..., n\}$.

The dynamics of the urn are given by its law function, which is as follows. For a possible history $\boldsymbol{x}$, let $k$ be the total number of membrane crossings in $\boldsymbol{x}$. For instance, if $N = 3$, and particles 1, 2, and 3 pass through the membrane 5, 1, and 3 times (respectively) during $\boldsymbol{x}$, then $k = 5 + 1 + 3 = 9$. Let $p$ be some small number, such as 0.01. Then

$$L_{\mathbf{X}}(\boldsymbol{x}) = p^k(1 - p)^{nN-k}.$$

Intuitively this law function means that each particle has chance $p$ of crossing to the other compartment in each unit of time, and chance $1 - p$ of staying where it is. It also says that the particles are stochastically independent. Since the quantity $p$ is independent of time, the law function is time-translation invariant. Also, the time reverse of a compartment switch is a compartment switch, so the law function is time-reversal invariant as well.

What is diffusion? Diffusion is a phenomenon that appears at the macro level, so let us define some macro-states for the urn. A macro-state of the urn is a specification of the number of particles in each compartment. We can use the notation $\langle a, b\rangle$ to represent the macro-state in which there are $a$ particles in compartment $A$ and $b$ in $B$. In this model, diffusion occurs when the numbers of particles in the two compartments become more nearly equal, such as in the transition from $\langle 15, 3\rangle$ to $\langle 11, 7\rangle$. A transition from $\langle 9, 9\rangle$ to $\langle 6, 12\rangle$, by contrast, is the opposite of diffusion.

As usual, I assume that the boundary condition is set at $t = 0$. We are then required to prove that there is a temporal asymmetry in $P_{\mathbf{X}}$ as far as diffusion is concerned. Roughly speaking, the system is disposed to undergo diffusion in the forward direction of time, and not in the backward direction. Of course, for diffusion to occur it must be physically possible. If the boundary condition is that the state at $t = 0$ is $\langle 9, 9\rangle$, then because this is already a state of equilibrium, or maximum diffusion, any further diffusion is impossible. We shall therefore assume that the boundary condition involves a macro-state like $\langle 15, 3\rangle$, which is far from equilibrium.

An important feature of a macro-state is the number of states that are consistent with it. (We may, if we wish, regard a macro-state simply as the class of states that are consistent with it.) The macro-state $\langle 0, 4\rangle$, for instance, has only one state, whereas the macro-state $\langle 1, 3\rangle$ is consistent with four states, as there are four particles that could be the one in compartment $A$. Let us call the logarithm of the cardinal number of a macro-state its *entropy*.[16] In general, for fixed $N$, the entropies of more diffused macro-states are much greater than those further from equilibrium.

The proof of forward diffusion relies on two facts: First, it uses Lemma 5.7.3.1, that *forward* transition chances are just the law values, or sums of law values, for sub-histories in that time interval. Second, it uses the fact that diffusion involves an increase of entropy.

A helpful way to proceed is to consider chances of single transitions, that is, forward transition chances of the form $P_{\mathbf{X}}(\mathbf{X}(t+1) = x' \mid \mathbf{X}(t) = x)$, and backward transition chances of the form $P_{\mathbf{X}}(\mathbf{X}(t-1) = x' \mid \mathbf{X}(t) = x)$. Using Lemma 5.7.3.1, the forward transition chance is just the law value of the sub-history $\boldsymbol{x}_t$ with state $x$ at $t$ and $x'$ at $t + 1$. If this sub-history $\boldsymbol{x}_t$ contains $j$ particle movements, then it may be shown that

$$L_t(\boldsymbol{x}_t) = p^j(1-p)^{N-j}.$$

We then have that

$$P_{\mathbf{X}}(\mathbf{X}(t+1) = x' \mid \mathbf{X}(t) = x) = p^j(1-p)^{N-j}.$$

Now we must consider how the value of this expression depends on the entropy change in the sub-history $\boldsymbol{x}_t$. The answer is, Not at all! Since $L_{\mathbf{X}}$ is tri, the time-reverse of $\boldsymbol{x}_t$ has exactly the same law value as $\boldsymbol{x}_t$, and so $P_{\mathbf{X}}(\mathbf{X}(t+1) = x \mid \mathbf{X}(t) = x') = P_{\mathbf{X}}(\mathbf{X}(t+1) = x' \mid \mathbf{X}(t) = x)$.[17] It may seem to follow from this that, in the interval $[t, t+1]$, the entropy is just as likely to go down as up. This is not the case, however, since if $x$ is a non-equilibrium state, then there is simply a *greater number* of sub-histories with increasing entropy than with decreasing entropy.

This may be shown with a simple example. Suppose $N = 4$, and the state $x$ is $\langle\{0\}, \{1,2,3\}\rangle$, that is, particle no. 0 is in $A$ and particles no. 1, 2, and 3 are in compartment $B$. This is a non-equilibrium state, as the equilibrium states are all members of the macro-state $\langle 2, 2\rangle$ and this is a member of $\langle 1, 3\rangle$. If the state $x'$ is $\langle 0, 4\rangle$ or $\langle 4, 0\rangle$, then entropy decreases. If $x'$ is a member of $\langle 1, 3\rangle$ or $\langle 3, 1\rangle$, then the entropy is constant. If the state $x'$ is in $\langle 2, 2\rangle$, then entropy increases. Table 5.1 displays all the important histories with their law values. (I neglect histories whose law values are less than $p^2(1-p)^2$ since $p \approx 0$, and so these have a negligible impact.)

TABLE 5.1

| From | To | Entropy | Law value |
|---|---|---|---|
| {0} {123} | { } {0123} | ↓ | $p(1-p)^3$ |
| {0} {123} | {0} {123} | – | $(1-p)^4$ |
| {0} {123} | {012} {3} | – | $p^2(1-p)^2$ |
| {0} {123} | {013} {2} | – | $p^2(1-p)^2$ |
| {0} {123} | {023} {1} | – | $p^2(1-p)^2$ |
| {0} {123} | {01} {23} | ↑ | $p(1-p)^3$ |
| {0} {123} | {02} {13} | ↑ | $p(1-p)^3$ |
| {0} {123} | {03} {12} | ↑ | $p(1-p)^3$ |

We see that there is one history with decreasing entropy, four with constant entropy, and three with increasing entropy. Overall, it is most likely that the entropy is constant, but an increase is about three times as probable as a decrease. (With a larger value for $N$ the chance of constant entropy is greatly reduced of course.)

When $N$ is large it is not difficult to see that, on each forward transition, about proportion $p$ of $A$'s particles will move to $B$, and about proportion $p$ of $B$'s particles will move to $A$. Thus, the overall flow will be about proportion $p$ of the difference between the numbers in $A$ and $B$. Now, this situation holds for every forward transition, so that in the history as a whole there will probably be a gradual evolution toward equilibrium in forward time.

Can this same reasoning be applied to the backward transition chances? Table 5.1 could certainly be switched around without any change in the law values, as the law function is time-reversal invariant. The argument stops there, however, since the backward transition chances are not equal to the corresponding law values. The reasoning would apply, at least to a reasonable approximation, for times long after the boundary condition, but in this region the system probably spends most of its time at equilibrium anyway, so there is no temporal asymmetry of diffusion. The temporal asymmetry of diffusion is a phenomenon that only exists near the boundary condition.

It should be stressed that this reasoning also applies to the time interval $t < 0$, if the system exists there. Thus, for negative times, the entropy will 'increase into the past,' that is, actually *decrease* over the interval $[-\tau, 0]$. This emphasizes the fact that the temporal asymmetry brought about by the boundary condition is local rather than global. In the interval $[-\tau, 0]$, the forward direction of time (i.e., the direction pointing away from the boundary condition) is the negative direction, towards $-\infty$.

We finally come to the fourth temporal asymmetry of chance, which is Reichenbach's Common Cause Principle (1963: 157–63). This principle (abbreviated as CCP) has been interpreted and used in a number of different ways, from being a way of *defining* causation to providing an argument for scientific realism. I will treat it as describing an empirically attested physical fact that requires a physical explanation. This is basically the way Reichenbach originally viewed it, I think. Its truth as a physical claim has been questioned by van Fraassen (1980: 28–31), particularly on the basis of observed correlations violating the Bell inequality. I agree that CCP is not universally valid, but it does hold within CSM. The discrepancy here is due to CSM3, which, though not universally valid, holds as a good approximation for 'large' systems. Van Fraassen also claims that CCP will fail in 'almost any indeterministic theory of sufficient complexity' (1980: 29), but this is doubtful since CCP is true within CSM, which allows for highly complex systems.

CCP is easily derived within CSM, unlike any other formalism I know of. It arises automatically, without any need for special, ad hoc, assumptions. Before we show this, however, we should say what CCP is.

Systems that do not directly interact, at least not significantly, often have actual histories that are frequency correlated. Moreover, I think, two such non-interacting systems **X** and **Y** may be correlated with respect to chance, that is,

$$P_{\mathbf{Z}}(\mathbf{X} \in B \mid \sigma(\mathbf{Y})) \neq P_{\mathbf{Z}}(\mathbf{X} \in B)$$
(in a set of non-zero probability), for some Borel set $B$.

A simple example of this is of two barometers in the same town. They will vary in sympathy, even though they do not interact with each other. In such cases of correlation without interaction, CCP says that there is always some other system, say **W**, that does interact with both **X** and **Y**. In our barometer example, the third stochastic process is the air pressure, of course.

There could be many systems like **W** that interact with both **X** and **Y**, but let us assume that, in this case, **W** is the only one.[18] CCP says that if we condition on the state of the common cause, then the correlation disappears, so we have

$$P_{\mathbf{Z}}(\mathbf{X} \in B_1 \ \&\ \mathbf{Y} \in B_2 \mid \sigma(\mathbf{W})) = P_{\mathbf{Z}}(\mathbf{X} \in B_1 \mid \sigma(\mathbf{W})).P_{\mathbf{Z}}(\mathbf{Y} \in B_2 \mid \sigma(\mathbf{W})), \text{ a.s.}$$

CCP involves a temporal asymmetry. To see this, let us suppose that **W** interacts with **X** and **Y** only in some bounded interval of time $T_i$. CCP then says that the **X**–**Y** correlation may exist after $T_i$, but not before. In other words, it may be that

$$P_{\mathbf{Z}}(\mathbf{X}_t \in B \mid \sigma(\mathbf{Y}_t)) \neq P_{\mathbf{Z}}(\mathbf{X}_t \in B),$$

but only if $t$ is after $T_i$. Before the interaction there can be no correlation at all. When Reichenbach says that conjunctive forks are always closed to the past, but usually open to the future, this is basically what he means.

The Common Cause Principle can therefore be expressed as the following four claims:

I If **X** and **Y** cannot interact, and there is no 'common cause' **W**, then **X** and **Y** are uncorrelated.

II Where there is such a system **W** that may interact with them both, **X** and **Y** can be correlated.

III In such cases of correlation, the correlation disappears when we condition on (the sigma field generated by) **W**.

IV If the interaction is restricted to some bounded interval of time, the correlation only exists after the interaction.

The first claim is just Theorem 5.4.1. II and III are easily deduced from the following theorem.

**5.7.3.5 Theorem** Let $\mathbf{Z} = \langle \mathbf{W}, \mathbf{X}, \mathbf{Y} \rangle$. If $bc_{\mathbf{Z}}$ entails that **X** and **Y** cannot directly interact, then $P_{\mathbf{Z}}(x \,\&\, y \mid w) = P_{\mathbf{Z}}(x \mid w) P_{\mathbf{Z}}(y \mid w)$.

*Proof*: Using Axiom 4 we have that $P_{\mathbf{Z}}(x \,\&\, y \mid w) = P_{\mathbf{Z}}(x \mid w) P_{\mathbf{Z}}(y \mid w \,\&\, x)$. By the causal theory of chance, this may be written as $Pr(x \mid \mathcal{L}_{\mathbf{Z}} \,\&\, bc_{\mathbf{Z}} \,\&\, w)$ $Pr(y \mid \mathcal{L}_{\mathbf{Z}} \,\&\, bc_{\mathbf{Z}} \,\&\, w \,\&\, x)$. The question here is whether $x$ is relevant to $y$ within the epistemic state $K = \mathcal{L}_{\mathbf{Z}} \,\&\, bc_{\mathbf{Z}} \,\&\, w$. Now, according to $\mathcal{L}_{\mathbf{Z}} \,\&\, bc_{\mathbf{Z}}$, **X** may interact with **W**, and **W** with **Y**, but **X** cannot interact directly with **Y**. If we let $I_{\langle \mathbf{W},\mathbf{X} \rangle}$ stand for '**W** may interact with **X**,' and so on, then the bridge between **X** and **Y** provided by $\mathcal{L}_{\mathbf{Z}} \,\&\, bc_{\mathbf{Z}}$ is the conjunction $I_{\langle \mathbf{W},\mathbf{X} \rangle} \,\&\, I_{\langle \mathbf{W},\mathbf{Y} \rangle}$. The proposition $\mathbf{W} = w$, however, gives complete information about the actual history of **W**, so using Corollary 5.4.2 the conjunctions $I_{\langle \mathbf{W},\mathbf{X} \rangle} \,\&\, w$, $I_{\langle \mathbf{W},\mathbf{Y} \rangle} \,\&\, w$ both factorize. Then, by Theorem 5.3.5, $x$ is not relevant to $y$ within the epistemic state $K = \mathcal{L}_{\mathbf{Z}} \,\&\, bc_{\mathbf{Z}} \,\&\, w$. The result is immediate.∎

**5.7.3.6 Corollary** If **X** and **Y** can each interact with **W**, though not with each other, then **X** and **Y** may be correlated.

*Proof*: Since **X** and **Y** can each interact with **W**, $P_{\mathbf{Z}}(x \mid w)$ and $P_{\mathbf{Z}}(y \mid w)$ can vary with $w$. Thus, since

$$\begin{aligned} P_{\mathbf{Z}}(x \,\&\, y) &= \sum_{w} P_{\mathbf{Z}}(x \,\&\, y|w) P_{\mathbf{Z}}(w) \\ &= \sum_{w} P_{\mathbf{Z}}(x|w) P_{\mathbf{Z}}(y|w) P_{\mathbf{Z}}(w), \end{aligned}$$

and neither term $P_{\mathbf{Z}}(x \mid w)$ nor $P_{\mathbf{Z}}(y \mid w)$ can be taken out of the sum as a constant, it may be that $P_{\mathbf{Z}}(x \,\&\, y \mid w) \neq P_{\mathbf{Z}}(x \mid w) P_{\mathbf{Z}}(y \mid w)$.∎

Note that to prove the existence, rather than the possibility, of a correlation would require consideration of a particular system, with specific dynamical properties. We thus have proved both II and III.

The final theorem is immediate, provided that we again interpret 'before' and 'after' in terms of closer to, and further from, the boundary condition. For, by Theorem 5.7.3.4, at times before the interaction $P_{\mathbf{X}}$ is exactly what it would be if there were no interaction with **W**, and the same is true of $P_{\mathbf{Y}}$. But, by Theorem 5.4.3, where there is no interaction there is no correlation. It follows then that there is no correlation between **X** and **Y** before the interactions, which is the result.

# 6

# Correlation

The topic of probabilistic correlation has become important, and puzzling, due to its central role in quantum theory. Einstein, Podolsky, and Rosen (1935) showed that the rules of quantum mechanics predict correlations between measurement results that seem to require 'hidden variables,' that is, physical quantities not represented in the quantum wave function. The hope of explaining these correlations using local hidden variables seemed to be quashed decisively by two results, however, one mathematical and one experimental. The mathematical result was Bell's theorem (Bell 1964), which showed, within certain assumptions, that any local hidden-variable theory must differ from quantum mechanics in some of its empirical predictions. The empirical result consisted of a number of different experiments to test the predictions of hidden-variable theories against those of quantum mechanics (QM), the best of which is considered to be that of Aspect et al. (1982). The verdict on these experiments, at least from 1980 onwards, is unanimous in favour of QM, and against hidden-variable theories of the type Bell considered.

The correlations discovered theoretically by Einstein, Podolsky, and Rosen (hereafter EPR), and experimentally confirmed by Aspect, are therefore a source of controversy at the present time. They have given rise to an extraordinary range of theories attempting to account for them. In this chapter we shall discuss the most promising explanations offered to date, and find them all unsatisfactory to various degrees.

We shall also see that the EPR correlations are provably impossible within CSM, which means that CSM cannot be generally valid; at best, it holds within some special case. It will be argued in section 6.3 that the postulate CSM3, though approximately true of 'large' systems, is false in general; and it will be shown that when this assumption is relaxed the inconsistency between

CSM and QM disappears. This result shows that a kind of 'hidden variable' explanation of EPR correlations is possible, within a stochastic framework. The hidden variables are not, of course, of the type Bell considered, nor are they non-local as in Bohm's theory (Bohm 1952).

## 6.1 CLASSICAL AND QUANTUM CORRELATION

The EPR correlations are correlations with respect to chance, but we shall see that they are quite different from the chance correlations allowed for within CSM. In a sense defined below, the correlations of CSM are classical, whereas the EPR correlations are not classical.

### *6.1.1 Classical Chance Correlations*

Let us review the results obtained in chapter 5 about chance correlations. First, systems that are necessarily causally independent at all times are uncorrelated (Theorem 5.4.3). Second, if systems **X** and **Y** possibly interact within the time interval $T_i$, then **X** and **Y** are correlated after $T_i$ but not before. What does it mean, exactly, to say that '**X** and **Y** are correlated after $T_i$'? Here is a suitable definition.

**6.1.1.1 Definition** **X** and **Y** are chance correlated at time t just in case the random variables **X**($t$) and **Y**($t$) are correlated w.r.t. chance.

By 'chance' here I mean the chance function $P_{\mathbf{Z}}$, where $\mathbf{Z} = \langle \mathbf{X}, \mathbf{Y} \rangle$. In Definition 5.6.9 I also defined a time-dependent chance function $P_t$, which is intuitively the chance at time $t$. One might wonder therefore which events are correlated with respect to $P_t$, and for which values of $t$. We find that, within CSM, time-slice subsystems are only correlated with each other if those time slices themselves interact. We have, in other words, the following theorem.

**6.1.1.2 Theorem** Suppose **X**, **Y** cannot interact outside $T_i$. Then, for $t^* > t > T_i$, **X**($t^*$) and **Y**($t^*$) are uncorrelated with respect to $P_t$.

*Proof*: The time slices of **X**, **Y** in $[t, \tau]$ do not interact, so we can apply Thm. 5.4.3.■

Since correlations within CSM all have this property, we shall call such correlations *classical* correlations.

**6.1.1.3 Definition** Suppose **X** and **Y** may interact within $T_i$ only. Then they are *classically* correlated (w.r.t. $P_Z$) iff **X** and **Y** are correlated w.r.t. $P_Z$, but uncorrelated w.r.t. $P_t$, where $t$ is any time after $T_i$.

One may be tempted to think that all chance correlations in nature are classical, but this is not so. We shall see that the formalism of quantum mechanics permits correlations that are not classical.

### *6.1.2 The Rules of Quantum Mechanics*

Quantum mechanics (QM) is a stochastic mechanical formalism, like CSM, but is set up rather differently. In this section I shall describe those of its properties that I need to appeal to later in the chapter. In QM a system **X** has a set of possible states, as in CSM. Each possible state is represented by a mathematical structure known as a *wavefunction* $\psi$. The idea that material particles such as electrons are associated with waves was first proposed by De Broglie (1924), and Schrödinger (1926) discovered a linear, deterministic equation that seemed to govern these 'matter waves.' The wavefunction for an electron, say, is a complex-valued function whose domain is the set of spacetime points, if we ignore electron spin. Schrödinger hoped that by means of this wavefunction, which at first he considered to represent some straightforward physical property, the discontinuities of Bohr's 'old' quantum theory could be eliminated. Schrödinger was able, for instance, to derive the Rydberg formula for spectral lines in the hydrogen atom from his wave equation.

The attempt to remove discontinuous changes from quantum theory failed, however, since Born showed that the intensity of the wavefunction has to be interpreted as a probability. Roughly speaking, for any position x, the intensity of the wavefunction at $x$, that is, $|\psi(x)|^2$, is the probability that the electron's position, if measured, will be found to be $x$. Thus, upon measurement, the state of a system evolves stochastically, rather than in accordance with Schrödinger's wave equation.

For a single particle, the domain of the wavefunction is just $\mathbb{R}^3$, which is the position space (space of possible positions) of a single particle. For a pair of particles, however, the wavefunction is defined on $\mathbb{R}^6$, as the two particles have six degrees of freedom in their positions. In general we might say that the domain of the wavefunction is the *configuration* space of the system concerned, although we must remember that the configuration space of a system in quantum mechanics is generally only half the size of the phase space of its classical counterpart. In classical mechanics, for example, the state of a particle is specified by its momentum as well as its position – a total of six numbers.

To describe the formalism as simply as possible we shall use the vector-space notation. A crucial fact about the set $H$ of possible wavefunctions for a system is that they form a vector space. That is, if $\psi_1, \psi_2 \in H$, then $\psi_1 + \psi_2 \in H$, and if $\psi \in H$ then $c\psi \in H$, where $c$ is any complex number. We can therefore consider each wavefunction in $H$ to be a *vector*; indeed, they are known as *state vectors*. In future we will denote $\psi$ in the 'ket' vector notation as $|\psi\rangle$.

We can define the scalar product of two ket vectors $|\psi_1\rangle$ and $|\psi_2\rangle$, written $g(|\psi_1\rangle, |\psi_2\rangle)$, as follows:

$$g(|\psi_1\rangle, |\psi_2\rangle) = \int \psi_1{}^* \psi_2 d\tau,$$

where $\tau$ ranges over the configuration space of $\psi$, and $c^*$ is the complex conjugate of $c$.

It is possible to define a second vector space, in addition to the ket vector space, by considering linear functions from ket vectors to complex numbers. In other words, we can consider functions $f$ from $H$ to $\mathbb{C}$, such that $f(|\psi_1\rangle + |\psi_2\rangle) = f(|\psi_1\rangle) + f(|\psi_2\rangle)$. It is trivial to show that the class of such functions also forms a vector space. Moreover, for any ket vector $|\psi\rangle$, we can associate it with a linear function $f_\psi$, using the metric $g$, as follows.[1]

**6.1.2.1 Definition** $f_\psi$ is the unique linear function $f$ such that $f(|\phi\rangle) = g(|\psi\rangle, |\phi\rangle)$, for all $|\phi\rangle \in H$.

Since $f_\psi$ is also a vector, we write it in the 'bra' vector notation as $\langle\psi|$. The space of linear functions on ket vectors is then known as the bra vector space. Using this bra-ket notation, the metric $g(|\psi\rangle, |\phi\rangle)$, which is equal to $f_\psi(|\phi\rangle)$, by definition, is written $\langle\psi \mid \phi\rangle$. Since $g$ has the form of an inner product, we can define the length or *norm* of a state vector $|\psi\rangle$ as $\langle\psi \mid \psi\rangle$, and speak of two state vectors $|\phi\rangle$ and $|\psi\rangle$ as *orthogonal* when $\langle\phi \mid \psi\rangle = 0$. We will assume that all ket vectors $|\psi\rangle$ are normalized, so that $|\langle\psi \mid \psi\rangle|^2 = 1$

The state vector $|\psi\rangle$ evolves in accordance with Schrödinger's equation when the system is isolated. When a measurement occurs, however, the situation is quite different. Each type of measurement on the system is represented by a linear 'operator' on $H$, which is a function $H \rightarrow H$. Each such operator $A$ has *eigenstates*, that is, states $|\psi\rangle$ such that $A|\psi\rangle = a|\psi\rangle$, where $a$ is a (complex) scalar called the *eigenvalue* corresponding to the eigenstate $|\psi\rangle$. A linear operator $A$ is called *Hermitian* if $g(|\phi\rangle, A|\psi\rangle) = g(A|\phi\rangle, |\psi\rangle)$. Hermitian operators have two important properties: (1) their eigenvalues are real, that is, their imaginary part is zero, and (2) the eigenstates corresponding to distinct eigenvalues are orthogonal.

The rule for measurement in quantum mechanics, known as the Projection Postulate, can now be expressed as follows. If a measurement is made that corresponds to a Hermitian operator $A$, then the possible states of the system after the measurement are just the eigenstates of $A$, which we can write as $|a_1\rangle$, $|a_2\rangle, \ldots, |a_i\rangle, \ldots$, regardless of the original state $|\psi\rangle$ of the system. The 'probability' of each eigenstate $|a_i\rangle$ is the squared scalar product or projection $|\langle a_i \mid \psi\rangle|^2$. It seems that 'probability' means the 'up to date' (utd) chance $P_t$, where $t$ is any time just before the measurement occurs. For each possible final state $|a_i\rangle$ the outcome of the measurement, that is, the measured value for the quantity associated with $A$ – is just $a_i$, the eigenvalue associated with $|a_i\rangle$.

There are operators for each classical property of the system, such as position, momentum, energy, angular momentum, and so on, as well as some new properties such as intrinsic spin. In general, however, a state vector does not contain a definite value for each of these properties. To say that a system possesses a definite value $a$ for quantity $A$ means, at the least, that a measurement of $A$ would necessarily yield the value $a$; yet for a quantum state $|\psi\rangle$ this only holds in the rare case that $|\psi\rangle$ is an eigenstate of $A$. The usual situation is that $|\psi\rangle$ is a linear combination, or superposition, of $A$'s eigenstates, that is, $|\psi\rangle = c_1 \mid a_1\rangle + c_2 \mid a_2\rangle + \ldots + c_i \mid a_i\rangle + \ldots$, where $\sum |c_i|^2 = 1$. If the property $A$ is measured on a system in such a superposition with respect to $A$, the wavefunction jumps to one of the eigenstates $|a_i\rangle$, as stated above. This change is sometimes called the 'collapse' of the wavefunction.

For two Hermitian operators $A$ and $B$ we define the *commutator* of $A$ and $B$, written $[A, B]$, as $AB - BA$. If $[A, B] = 0$, that is, $[A, B]|\psi\rangle = 0$ for all $\psi \in H$, then the operators $A$ and $B$ are said to *commute*. If $A$ and $B$ commute, then it may be shown that they share the same set of eigenstates, which means that a single state vector may determine a precise value for both $A$ and $B$. If $A$ and $B$ do not commute, then there is no state vector for which $A$ and $B$ both have precise values.

Since, for each state $\psi$, there is a chance distribution over the eigenvalues of $A$, there are degrees to which a quantity $A$ has a value in state $\psi$. $A$ will have an expected value $\langle A\rangle$, defined in the usual way as the sum of its possible values multiplied by their probabilities, from which it follows that $\langle A\rangle = \langle\psi|A|\psi\rangle$. The standard deviation of $A$, written $\Delta a$, can then be defined in the usual way as $\Delta a = \sqrt{(\langle A^2\rangle - \langle A\rangle^2)}$. The smaller the standard deviation $\Delta a$, the more precisely defined the quantity $A$ is. Indeed, when $\psi$ is an eigenstate of $A$, $\Delta a = 0$. The standard deviation of $A$ is referred to in QM (rather confusingly) as the *uncertainty* of $A$. One important relation in quantum mechanics, which is a theorem of the formalism, is known as the Heisenberg Uncertainty Relation. It is that, for any two operators $A$ and $B$,

$$\Delta a \Delta b \geq \tfrac{1}{2}|\langle [A, B] \rangle|$$

Thus, if $A$ and $B$ do not commute, then there is an upper limit to their simultaneous precision. If the uncertainty of one is reduced, then this forces the system into a state where the other is more uncertain, so that $\Delta a \Delta b$ does not drop below $\tfrac{1}{2}|\langle [A, B] \rangle|$. Either of $A$ or $B$ can be made to have an arbitrarily precise value, but only at the expense of the other. It turns out that canonically conjugate quantities, such as position and momentum, do not commute. Thus, at any time, a quantum system has definite values for at most half of the quantities that define a classical state.

The feature of quantum mechanics upon which we shall focus in this chapter is the way composite systems are dealt with. This is very elegant and obvious, but has the surprising consequence that non-classical correlations are possible. Let us consider the composite system $\mathbf{Z} = \langle \mathbf{X}, \mathbf{Y} \rangle$, where the Hilbert spaces of **X** and **Y** are $H_{\mathbf{X}}$ and $H_{\mathbf{Y}}$ respectively. The Hilbert space of $\langle \mathbf{X}, \mathbf{Y} \rangle$ is then just the tensor product $H_{\mathbf{Z}} = H_{\mathbf{X}} \otimes H_{\mathbf{Y}}$, as one might expect. Also, if **X** is in state $|\phi\rangle_{\mathbf{X}}$, and **Y** is in state $|\psi\rangle_{\mathbf{Y}}$, then **Z** is in state $|\phi\rangle_{\mathbf{X}} \otimes |\psi\rangle_{\mathbf{Y}}$, which we write $|\phi\rangle_{\mathbf{X}}|\psi\rangle_{\mathbf{Y}}$. This seems very mundane, but now let us suppose that **X** and **Y** interact, in the time interval [3, 4] say. During the interaction the state of **Z** evolves in accordance with Schrödinger's equation, ending up in a state $|\Psi\rangle_{\mathbf{Z}}$ say at $t = 4$. We would perhaps expect that $|\Psi\rangle_{\mathbf{Z}}$ somehow specifies an independent state for the subsystem **X**, and also one for **Y**, but this is not so. It would only be this way if $|\Psi\rangle_{\mathbf{Z}}$ could be 'factorized' into a tensor product of vectors in $H_{\mathbf{X}}$ and $H_{\mathbf{Y}}$, that is, be expressed as $|\Psi\rangle_{\mathbf{Z}} = |\phi\rangle_{\mathbf{X}}|\psi\rangle_{\mathbf{Y}}$, which is not the case in general. $|\Psi\rangle_{\mathbf{Z}}$ can always be expressed as a linear combination of such products, but not always as a single product.

What this means is that, after **X** and **Y** interact with each other, they each cease to have a state vector. The composite system **Z** continues to have a state vector, as it is a closed system, but its component parts do not. Nothing like this happens within CSM, where interacting systems retain their individual states at all times. As we shall see below in the discussion of the EPR argument, two quantum systems that have lost their individual states in this way are non-classically correlated.

It is possible to represent a quantum state using an operator on the Hilbert space. One may wonder about the value of this, but we see below that it is very useful. The operator on $H$ we need to represent the wavefunction $\psi$ is the one that maps a ket vector $x$ to $f_{\psi}(x).|\psi\rangle$, that is, it maps $|\phi\rangle$ to $\langle \psi \mid \phi \rangle.|\psi\rangle$. This operator is known as the *projection operator*, or just *projector*, for the state $|\psi\rangle$. Intuitively, it maps each $|\phi\rangle$ onto a vector parallel to $|\psi\rangle$, whose length is the projection of $|\phi\rangle$ onto $|\psi\rangle$. The projector for $\psi$ is usually written $|\psi\rangle\langle\psi|$.

Given a particular orthonormal basis $|e_i\rangle$ for $H$, any operator $A$ on the Hilbert space may be expressed as the matrix of complex numbers $A_{rs} = \langle e_r|A|e_s\rangle$. In the particular case of a projector $|\psi\rangle\langle\psi|$, this entails that the operator can be expressed as a matrix $P_{rs}$, as follows.

$$\text{If } \psi = \sum_i c_i|e_i\rangle, \text{ then } P_{rs} = c_r c_s^*.$$

We then define the *trace* of $P$, written $TrP$, as follows:

$$TrP = \sum_i P_{ii},$$

so that the trace is just the sum of the diagonal elements of the matrix. (It may be shown that the trace of an operator is independent of the orthonormal basis chosen.) It is now possible to calculate the probabilities of measurement outcomes using the projectors. If a system is in state $|\psi\rangle$, and we measure an operator $A$, then the probability of the outcome $a$ is just $Tr(P_\psi P_a)$, where $P_\psi = |\psi\rangle\langle\psi|$ and $P_a = |a\rangle\langle a|$.

So we see that, in a sense, the operator $|\psi\rangle\langle\psi|$ is an alternative representation for the state $|\psi\rangle$. What advantage is there in this representation, however? Suppose a physicist is unsure of the actual state of a system **X**; it could be in any of the states $|\psi_1\rangle$, $|\psi_2\rangle$, $|\psi_3\rangle$, and so on, with epistemic probabilities $p_1$, $p_2$, $p_3$, etc. Let us now define the following operator $\rho$, which is called a *density operator*.

$$\rho = \sum_i p_i|\psi_i\rangle\langle\psi_i|.$$

The density operator has the nice property that it delivers *epistemic* probabilities of measurement outcomes, namely: $P_K(A = a) = Tr(\rho P_a)$. It may also be shown that the *epistemic* expectation of an observable $A$ in state $\rho$ is given by $Tr(\rho A)$. The density operator is sometimes described as a (classical) *mixture* of projectors, as it is simply a weighted mean. A state represented by a density operator[2] is therefore called a *mixed* state, in contrast to the *pure* state represented by a projector.

Is it possible to distinguish experimentally between a pure state and a mixture? More specifically, can one tell the difference between the superposition of states

$$|\psi\rangle = \sum_i c_i|a_i\rangle,$$

and the mixture of those very same states? (i.e., $\rho = \Sigma_i\, p_i|a_i\rangle\langle a_i|$, where $p_i = |c_i|^2$). In fact it is indeed possible, as I shall now explain.

Clearly, if the states $\{|a_i\rangle\}$ are eigenfunctions of the observable $A$, then measuring $A$ on $|\psi\rangle$ and on $\rho$ give identical statistics. Thus, one cannot distinguish $|\psi\rangle$ from $\rho$ by measuring $A$. Let us therefore measure another observable $B$ that does not commute with $A$, so that it has different eigenstates. If **X** is in state $|\psi\rangle$, then the chance of the outcome $B = b$ is

$$P(B = b) = \left|\sum_k c_k\langle b|a_k\rangle\right|^2. \qquad (1)$$

For the state $\rho$, however (i.e., the system is really in *one* of the states $\{|a_i\rangle\}$, each with epistemic probability $|c_i|^2$), we have

$$P_K(B = b) = \sum_k |\langle b|a_k\rangle|^2|c_k|^2. \qquad (2)$$

The important difference between these two expressions is that in (1) the terms in the summation are complex, whereas in (2) they are real. Thus, in the pure state $\psi$ there will, in general, be interference between the elements of the superposition, but for a mixture this is impossible.

To clarify the situation, let us consider a third possible state of knowledge, where we know that the system is in some superposition $\psi$ of eigenvectors $\{|a_k\rangle\}$, with coefficients $c_k$, but we do not know the *phases* of the $c_k$. In other words, if we write $c_k = |c_k| \exp(i\theta_k)$, then we are ignorant of the phases $\theta_k$. This being so, in calculating $P_K(B = b)$ we must treat each $\theta_k$ as a variable and integrate the expression (1) separately w.r.t. each unknown phase. This yields

$$\begin{aligned} P_K(B = b) &= \int\left|\sum_k |c_k|\exp(i\theta_k)\langle b|a_k\rangle\right|^2 . \, dP(\theta_k) \\ &= \sum_k |\langle b|a_k\rangle|^2|c_k|^2, \qquad (3) \end{aligned}$$

assuming that the phases each have a uniform probability distribution, and are pairwise probabilistically independent. Thus, the statistics in case (3) are identical to those in (2).

In the third case the system is in a superposition of states, but the phases of the coefficients are unknown. Now, a superposition is often described as *coherent*, which means that the relative phases of the elements are determinate and precise. It is this property of coherence that gives rise to the famous interference phenomena of quantum mechanics, such as in the double-slit experiment.[3] In the epistemic state of ignorance about the phases, however, the elements are incoherent, in the sense that the relative phases are unknown or indeterminate. Since this epistemic state is equivalent to a mixture, mixtures are often described as incoherent.

It should be noted that the introduction of density operators is merely a formal convenience, and does not alter the theory itself (it is a conservative extension, in other words). It does, however, enable some features of the formalism to become clear. Two such features, which are considered significant, will now be briefly discussed. First, when two systems are correlated, so that neither has a state vector, each system can still be assigned a density operator. Second, the decomposition of a density operator into projectors is not unique. It is always possible to expand a density operator in two or more alternative orthonormal bases.

Suppose **X** and **Y** are correlated. The state vector of $\mathbf{Z} = \langle \mathbf{X}, \mathbf{Y} \rangle$ may then be expressed as the following sum

$$|\Psi\rangle_{\mathbf{Z}} = \sum_{r,s} c_{rs} |\varphi_r\rangle_{\mathbf{X}} |\psi_s\rangle_{\mathbf{Y}},$$

where $|\phi i\rangle_{\mathbf{X}}$ and $|\psi_i\rangle_{\mathbf{Y}}$ are orthonormal bases for **X** and **Y** respectively. Now consider an observable $A$ for the system **X** only – perhaps $A$ is the operator for a measurement of the momentum of **X**. If we want to calculate the probabilities of measurement outcomes of $A$ on **Z** in state $|\Psi\rangle_{\mathbf{Z}}$, we must apply $A$ to vectors in $H_{\mathbf{Z}}$ rather than $H_{\mathbf{X}}$. For a vector in $H_{\mathbf{Z}}$ of the product form $|\phi\rangle_{\mathbf{X}}|\psi\rangle_{\mathbf{Y}}$ this is straightforward: the image under $A$ is just $(A|\phi\rangle_{\mathbf{X}})|\psi\rangle_{\mathbf{Y}}$. In other words, we apply $A$ only to the factor that exists in $H_{\mathbf{X}}$. To apply $A$ to a general vector $|\Psi\rangle_{\mathbf{Z}}$ in $H_{\mathbf{Z}}$ we express $|\Psi\rangle_{\mathbf{Z}}$ as a sum of such products, apply $A$ to each product, and then sum up the images (which is valid because $A$ is linear).

As with a single system, it is convenient to express operators as matrices. Previously we expressed a ket vector as a column vector, but it is simpler to represent the ket vector $|\Psi\rangle_{\mathbf{Z}}$ as the *matrix* $c_{rs}$. The reason for this is that each column of this matrix can be considered as a column vector, and in each such vector the base vector for **Y** is constant. Thus, using the rule above, we can apply $A$ to $|\Psi\rangle_{\mathbf{Z}}$ using the matrix product, namely:

$$(A|\Psi\rangle)_{rs} = \sum_i A_{ri}c_{is}.$$

How can we calculate the value of a scalar product $\langle\Phi \mid \Psi\rangle$ of two vectors in $H_{\mathbf{Z}}$ using this matrix notation? If $|\Phi\rangle_{\mathbf{Z}}$ is $d_{rs}$ and $|\Psi\rangle_{\mathbf{Z}}$ is $c_{rs}$, then

$$\langle\Phi \mid \Psi\rangle = \sum_{r,s} c^*_{rs}d_{rs}.$$

Using these facts, let us calculate the expectation of $A$ when **Z** is in state $|\Psi\rangle_{\mathbf{Z}}$. We have:

$$\begin{aligned}\langle A\rangle &= \langle\Psi|A|\Psi\rangle \\ &= \langle\Psi| \sum_i A_{ri}c_{is} \\ &= \sum_{r,s,i} c^*_{rs}A_{ri}c_{is} \\ &= \sum_{r,s,i} c^*_{rs}c_{is}A_{ri}.\end{aligned}$$

This quantity is just the trace of the matrix $\rho^{\mathbf{X}}A$, where

$$\rho_{rs}{}^{\mathbf{X}} = \sum_i c_{ri}c^*_{si},$$

so that $\rho^{\mathbf{X}}$ is the density matrix for the system **X**.[4]

Let us now consider some arbitrary element $|\psi_k\rangle_{\mathbf{Y}}$ of the orthonormal basis for **Y**. Following Everett, we can now define the *relative* state $|\xi_k\rangle_{\mathbf{X}}$ for the system **X**, w.r.t. the state $|\psi_k\rangle_{\mathbf{Y}}$ for **Y**, within $|\Psi\rangle_{\mathbf{Z}}$, as follows.

$$|\xi_k\rangle_{\mathbf{X}} = M_k \sum_i c_{ik}|\varphi_i\rangle_{\mathbf{X}},$$

where $M_k$ is just a normalization factor. This is just the $k^{\text{th}}$ column of the matrix of coefficients for $|\Psi\rangle_{\mathbf{Z}}$, as this is the column containing all and only coefficients of terms like $|\phi_i\rangle_{\mathbf{X}}|\psi_k\rangle_{\mathbf{Y}}$. We can now simplify the expression of $|\Psi\rangle_{\mathbf{Z}}$, as

$$|\Psi\rangle_{\mathbf{Z}} = \sum_j \frac{1}{M_j}|\xi_j\rangle_{\mathbf{X}}|\psi_j\rangle_{\mathbf{Y}}.$$

(Note the single variable of summation.) More importantly, however, the density operator $\rho^{\mathbf{X}}$ for **X** can now be expressed as

$$\rho^{\mathbf{X}} = \sum_i p_i|\xi_i\rangle\langle\xi_i|,$$

$$\text{where } p_i = \sum_i |c_{ji}|^2.$$

Thus, the system **X** behaves exactly as if it had a definite (but unknown) state $|\xi_k\rangle_{\mathbf{X}}$! If we are restricted to making measurements on **X**, rather than on both **X** and **Y**, then we cannot distinguish between this situation where **X** has no (pure) state and one where it has some unknown pure state $|\xi_k\rangle_{\mathbf{X}}$.

It should be noted that the orthonormal basis $\{|\psi_i\rangle_{\mathbf{Y}}\}$ is perfectly arbitrary, and so some other basis $\{|\psi'_i\rangle_{\mathbf{Y}}\}$ could be used instead. By the same reasoning we then infer that **X** behaves exactly as if it had some definite state in the set $\{|\xi'_i\rangle_{\mathbf{X}}\}$. But the two sets $\{|\xi_i\rangle_{\mathbf{X}}\}$ and $\{|\xi'_i\rangle_{\mathbf{X}}\}$ may be entirely disjoint! Thus, we cannot conclude that **X** *really* has a pure state that is some member of $\{|\xi_i\rangle_{\mathbf{X}}\}$, for then by parity of reasoning we would have to conclude that it also has a pure state in $\{|\xi'_i\rangle_{\mathbf{X}}\}$, and these two conclusions are sometimes inconsistent.

### *6.1.3 The Orthodox Interpretation*

In order to put EPR's argument into its historical context, I shall first briefly outline the interpretation of QM devised by some of its founders, which in 1935 was regarded as orthodox. This view was developed by a group of researchers clustered around Niels Bohr, based at the Institute of Theoretical Physics in Copenhagen, and so is known as the Copenhagen Interpretation.

The Copenhagen interpretation may be represented by the following five principles.[5]

1 Microscopic entities (such as electrons and atoms) do not possess in themselves values for physical quantities (except for conserved quantities such as mass and charge). These values are only defined in the context of, that is, relative to, a definite experimental arrangement, involving preparation and measuring devices.
2 There exists what Bohr repeatedly refers to as an 'unanalyzable link' between the microsystem and the macroscopic measurement apparatus.
3 Macroscopic experimental arrangements, and the results of experiments, are to be described in the language of classical physics.[6]

4 The principle of complementarity: it is not necessary (or possible) to construct a *single* description of a microsystem that is accurate across different experimental arrangements. Since different experimental arrangements exclude one another, the different perspectives they provide on the microsystem cannot contradict one another, but should be regarded as *complementary*. We must give up the classical idea of a unified, perspective-free description of reality when dealing with microsystems.
5 The formalism of quantum mechanics cannot therefore be made more complete by adding quantities that are not defined by measuring instruments. Quantum mechanics is not complete in any naive sense of corresponding exactly to a unified reality that is somehow 'given,' but does capture every quantity that is physically meaningful.

These principles all arise rather easily from the formalism of QM, as can be seen even from our cursory presentation in §6.1.2, although this will not be fully apparent until we come to discuss the EPR argument. The main point for now is that, within the formalism of QM, a statement like 'the electron has position $q$' can only be interpreted as 'the state vector of the electron is an eigenstate of the position operator $Q$, with eigenvalue $q$.' Thus, it appears, the very *meaning* of the ascription of position to the electron requires a reference to a measuring device, that is, an experimental arrangement.

The principle of complementarity is important in Bohr's understanding of the Heisenberg uncertainty principle. Suppose there were two non-commuting operators that were both measurable together, in the same experimental arrangement. This would cause a major difficulty for QM, since there is no quantum state in which both quantities have precise values simultaneously. We would have a contradiction within the theory. It turns out, however, that measurements of non-commuting quantities are always exclusive, in the sense that it is physically impossible to carry out precise measurements of both at one time. (This is not true of commuting pairs of operators, which can be measured together.) This fact that the measurements are exclusive is not a matter of luck; rather, the operators do not commute *because* the measurements are exclusive. It is only meaningful to ascribe precise values to two quantities together if it is possible to measure the two together.[7]

We will discuss the Copenhagen interpretation more fully in the next chapter, when we deal with the measurement problem, but to avoid misunderstanding a few remarks should be made at this stage about anti-realism. The Copenhagen interpretation has often been presented as a form of anti-realism, or relativism, or instrumentalism. Anti-realism is essentially the claim that there is no 'God's-eye view' of the world – no way that things actually stand, independent of all possible human knowledge. It collapses the distinction between justification and truth, and generally rejects metaphysics in favour of epistemology. A committed anti-realist does not even understand what it *means* to claim that a theory is true, rather than merely justified. Relativism, or perspectivism, seems in most contexts to be the same thing as anti-realism. It claims that the different perspectives belonging to different 'observers' (in the most general sense) cannot be understood as relations to a single reality. Rather the perspectives, or 'realities,' are all that exists. Instrumentalism is similar to these views, as it states that scientific theories (such as QM) should not aim at truth, or correspondence to objective reality, but should merely attempt to save the phenomena. To characterize Bohr's own philosophical ideas in such terms may well be accurate, and is certainly justified by many of his remarks, but I believe that the Copenhagen interpretation as presented here is not essentially wedded to anti-realism.

There is an important difference between the claim that there is no objective, perspective-independent reality and the claim that physical quantities of microsystems are only defined relative to a measurement apparatus. The essential point is that relations can be just as objective as properties, so that a discovery that a quality once thought to be a property turns out to be relational does not threaten objectivity. For instance, Newton thought that the time interval between two events was a property of those events, but we now think that this duration is only defined within a particular reference frame. This does not mean that there is no perspective-independent reality, no fact of the matter, but merely that *durations* are not part of that reality in the way we previously thought. In the next chapter I shall argue for an interpretation of QM that is similar to the Copenhagen interpretation, yet thoroughly realist.

### *6.1.4 The EPR Argument*

The 1935 paper by Einstein, Podolsky, and Rosen arose out of a long dialogue between Einstein and Niels Bohr[8] on the interpretation of quantum theory. The earlier thought experiments devised by Einstein were intended to show that, under certain conditions, it is possible to make simultaneous measurements of non-commuting quantities to a greater precision than is allowed by the Heisenberg uncertainty relation. He was motivated by the desire to show that QM as it stands is not a complete theory, but is instead like statistical mechanics, dealing with mass phenomena. He wanted to interpret the uncertainty of a physical quantity as the standard deviation of the quantity in an ideal ensemble of systems prepared in a similar way. In other words, Einstein wanted to show that quantities always had precise values, even if they could not be *measured* precisely.[9] Einstein was not overly worried by the indeterminism of QM, but rather with Bohr and Heisenberg's idea that types of measurement actually *define* physical quantities, so that the idea of a precise value that cannot be measured, even in principle, is absurd. Einstein wanted to preserve the realist's intuition that the world can outstrip our knowledge, or even possible knowledge, of it.

The EPR paper was based on their 'criterion of physical reality,' an idea of the greatest importance. It is as follows:

> If, without in any way disturbing a system, we can predict with certainty (i.e. with probability equal to unity) the value of a physical quantity, then there exists an element of physical reality corresponding to this physical quantity. (138)

If this principle is true, and we assume locality, then it does follow that QM is incomplete, by the following reasoning. Consider two systems **X** and **Y** that interact in some interval of time, say [3, 4]. After $t = 4$, the individual systems no longer have state vectors, as pointed out above, although $\mathbf{Z} = \langle \mathbf{X}, \mathbf{Y} \rangle$ still has the state vector $|\Psi\rangle_{\mathbf{Z}}$. If $A$ is the operator for some quantity on system **X**, then let its eigenstates be $|a_1\rangle, |a_2\rangle, |a_3\rangle, \ldots$ and the corresponding eigenvalues be $a_1, a_2, a_3, \ldots$ Now, according to the rules of QM, $|\Psi\rangle_{\mathbf{Z}}$ may be expressed as follows:

$$|\Psi\rangle_{\mathbf{Z}} = \sum c_i |a_i\rangle_{\mathbf{X}} |\varphi_i\rangle_{\mathbf{Y}},$$

where $|\varphi_i\rangle$ is the relative state for **Y** with respect to **X** being in state $|a_i\rangle$, when $\langle \mathbf{X}, \mathbf{Y} \rangle$ is in state $|\Psi\rangle$, as defined by Everett (1957: 317). Suppose $A$ is now measured on **X**, yielding the value $a_k$. This means that **X** now has the state vector $|a_k\rangle$, and **Y** also now has a state vector, namely, $|\varphi_k\rangle$. We are not forced, of course, to measure $A$ on **X**, as we could measure anything we like, such as $B$ instead. $B$ has eigenstates $|b_1\rangle, |b_2\rangle, |b_3\rangle, \ldots$ with eigenvalues $b_1, b_2, b_3, \ldots$ Let us then expand $|\Psi\rangle$ using the orthonormal basis defined by $|b_1\rangle, |b_2\rangle, |b_3\rangle, \ldots$ as follows:

$$|\Psi\rangle = \sum c_i |b_i\rangle |\psi_i\rangle,$$

where $|\psi_i\rangle$ is the relative state for **Y** w.r.t. **X** being in state $|b_i\rangle$, when $\langle \mathbf{X}, \mathbf{Y} \rangle$ is in state $|\Psi\rangle$. If we measure $B$ on **X** instead of $A$, **X** will be left in some eigenstate of $B$, such as $|b_s\rangle$, which means that **Y** will have the state vector $|\psi_s\rangle$ instead of $|\varphi_k\rangle$. Now, since **X** and **Y** are no longer interacting at the time of **X**'s measurement, the choice of measuring $A$ or $B$ cannot have any physical effect on **Y**. Thus, as EPR put it, it is possible to assign two different state vectors, $|\psi_s\rangle$ and $|\varphi_k\rangle$, to the same reality. The problem with this is that there are cases where the $\{|\varphi_k\rangle\}$ are the eigenstates of some operator $P$, and $\{|\psi_s\rangle\}$ of some operator $Q$, where $P$ and $Q$ do not commute. Let $|\varphi_k\rangle$ have eigenvalue $p$ for $P$, and $|\psi_s\rangle$ have $q$ for $Q$. Then, without in any way disturbing **Y**, we can (by *alternative* measurements on **X**) predict with certainty either the result of a measurement of $P$ or of $Q$ on **Y**. There must therefore be an element of physical reality in the system **Y** to determine the result of both a $P$-measurement and a $Q$-measurement, as **Y** cannot 'know' which measurement on **X** has been performed. It must hold answers to both questions in readiness, so to speak. But quantum mechanics does not represent these elements of reality; indeed, within QM it is not even possible for $P$ and $Q$ to have precise values simultaneously. Since, then, QM fails to represent some aspects of physical reality, it is incomplete.

The EPR argument is certainly valid, in that the conclusion does follow from the premisses. There are two premisses that have been doubted, however. First, one may doubt that two systems that have a combined state but no individual states can ever cease interacting, even if they are far apart in space. In short, one can deny the *locality* assumption of the argument. This is difficult to do, as the correlation between measurements on **X** and **Y** exists even if the measurement events are space-like separated, so that any signalling between them must be super-luminal. Second, one may doubt the reality criterion.

Let us look more closely at the reality criterion. It seems to involve two separate ideas about reality, both of which are very plausible. We note first that the criterion involves the notion of *prediction*. What is prediction? It is a matter of inferring the later behaviour of a system from some information about its earlier behaviour. Thus, when the criterion speaks of being able to predict the value of a physical quantity, we might wonder what information is used to make the prediction.

In the example to which the criterion is applied, the total system can be considered to consist of four subsystems – **X**, **Y**, and a couple of measuring instruments, $\mathbf{M}_B$ and $\mathbf{M}_Q$ say, which measure the quantities $B$ and $Q$. Now suppose we are able to predict with certainty the result of $Q$ being measured on **Y**, that $Q = q$ perhaps; this is really a prediction about the system $\langle \mathbf{Y}, \mathbf{M}_Q\rangle$. The question then arises as to which systems we are allowed to have knowledge of in making this prediction. The criterion itself suggests no restrictions here, and in the application of the criterion we use the result of a measurement on **X** to make such a prediction; thus, I assume, we are allowed knowledge of *all* systems, not just the system $\langle \mathbf{Y}, \mathbf{M}_Q\rangle$ that the prediction is about.

The first part of the reality criterion is then as follows: If, given knowledge of the world up to time $t$, we can predict with certainty the result of a measurement at $t + d$, then there exists an element of reality at $t$ which determines that outcome. This is practically a tautology. There is surely no possibility that it is false. If the value of $\mathbf{M}_Q(t + d)$ can be predicted with certainty from the states of $\langle \mathbf{X}, \mathbf{Y}, \mathbf{M}_B, \mathbf{M}_Q \rangle$ up until time $t$, this just means that there are some true propositions about the latter that entail, that is, determine, the value of $\mathbf{M}_Q(t + d)$.

One will note that this part of the criterion has nothing to say about not disturbing the system in question. What work does this part of the criterion do? Although the criterion itself is not explicit about this, I think it is a matter of where the 'element of reality' corresponding to the prediction is *located*. The second part of the criterion, I believe, is that if $\langle \mathbf{Y}, \mathbf{M}_Q \rangle$ is causally isolated from $t$ onwards, and its state at $t + d$ is predictable with certainty from the state of the world at t, then the element of reality corresponding to this prediction 'resides' in the system $\langle \mathbf{Y}, \mathbf{M}_Q \rangle$ itself. Without this additional claim about the location of the element of reality, the words 'without in any way disturbing the system' are unnecessary. We should note also that the EPR argument does not work if the element of reality is even partially located at **X**. For then the element of reality may depend upon which type of measurement ($A$ or $B$) is performed at **X**, in which case the elements of reality corresponding to the predictions $P = p$ and $Q = q$ need not both exist together. It is only if they are safely placed at **Y**, immune from **X**'s influence, that they are guaranteed both to exist. We may summarize these parts as follows.

**C1** If an event $E$ at $t_2$ in system **Y** can be predicted with certainty from the state of the world up until $t_1 < t_2$, then the world at $t_1$ contains some element of reality corresponding to $E$.

**C2** If **Y** is also causally isolated from $t_1$ onwards, then the element of reality resides in **Y**.

We should now investigate the meaning of 'element of reality.' The most charitable approach here is to take the weakest interpretation for which the EPR argument is valid; so let us look at how EPR actually use the term. We find that if a measurement result $A = a$ is *predictable* from the state of a system **X**, then **X** contains an element of reality corresponding to that prediction. An element of reality corresponding to a measurement result $A = a$, therefore, is a physical fact about some system from which the result $A = a$ may be predicted. It should be noted that EPR offer this merely as a sufficient, and not a necessary, condition of physical reality.

It may have been noticed that I am departing slightly from Einstein's terminology in speaking of an element of reality corresponding to a measurement prediction rather than a value of a physical quantity. My formulation is slightly more general than Einstein's, while leaving the basic idea unchanged. The point is that Bohr, to whom the argument was addressed, considered an assertion that a system **X** has value $a$ for some quantity $A$ to posit a relation between **X** and a measuring instrument. The value $a$ for the quantity $A$ does not exist within **X** alone, but only within the composite system $\langle \mathbf{X}, \mathbf{M}_A \rangle$. We can grant this to Bohr, at least for the sake of argument, and still get the same conclusion.

Using this sharpened reality criterion we can state the EPR argument as follows. Suppose **X** and **Y** do not interact after $t = 4$. At $t = 5$ we can measure either $A$ or $B$ on **X**, so let us consider two possible worlds: $w_A$ in which $A$ is measured on **X**, and $w_B$ where $B$ is measured; otherwise, $w_A$ and $w_B$ are as similar as they can be. In $w_A$, $A$ on **X** will have some definite value at $t = 5$, and so $P$ on **Y** will also be value-definite. In $w_B$, $B$ on **X** will have some definite value at $t = 5$, and so $Q$ on **Y** will also be value-definite. But, since **Y** is isolated from $t = 4$ onwards, $w_A$ and $w_B$ do not differ at all with respect to the system **Y**. As EPR put it, 'no real change can take place in the second system in consequence of anything that may be done in the first system' (1935: 140).

We now apply the reality criterion. By C1, there is some element of reality in $w_A$ at $t = 5$ corresponding to $P = p$ say, and in $w_B$ at $t = 5$ there is some element of reality corresponding to $Q = q$. Also, by C2, this element of reality resides in the system **Y** in each case. Now, since **Y** is unchanged between $w_A$ and $w_B$, these two elements of reality *both* exist in each world! Thus, it is possible for $P$ and $Q$ to possess values simultaneously, and so QM is incomplete.

### *6.1.5 Predictive and Causal Locality*

The reality criterion C2 is a kind of locality principle, although it is quite different from what one normally means by 'locality.' Locality is the view that causal interaction can only occur between events that are contiguous in spacetime – in short, there is no 'action at a distance.' Locality in the usual sense, therefore, is concerned with *causation*. The principle C2, by contrast, is concerned with prediction, or determination. The claim is not that the measurement result $Q = q$ is caused by $\langle \mathbf{Y}, \mathbf{M}_Q \rangle$ alone; this is trivially true, since $\langle \mathbf{Y}, \mathbf{M}_Q \rangle$ is by hypothesis a causally isolated system. Rather, C2 holds that the measurement result can be *predicted* from knowledge about $\langle \mathbf{Y}, \mathbf{M}_Q \rangle$ alone, so that facts about $\langle \mathbf{Y}, \mathbf{M}_Q \rangle$ are in themselves sufficient to *determine* the result $Q = q$.

We saw in chapter 3 that causation and determination are distinct relations, since causation is neither necessary nor sufficient for determination. Indeed, even the relata of the relations are different, as causation is concerned with concrete events while determination is defined on possible states of affairs. We must not confuse C2 with the trivial claim that, since **Y** is causally isolated, outcomes of measurements on **Y** are caused by **Y** alone.

The criterion C2 can be strengthened in an obvious way, to yield what we might call the principle of determinative locality, or *Predictive Locality*, as follows.

**Predictive Locality** Suppose a system **Y** is isolated from time $t_0$ onwards, and $t_0 < t_1 < t_2$. Then, in making predictions about $\mathbf{Y}(t_2)$ from the state of the world at $t_1$, only $\mathbf{Y}(t_1)$ is relevant. In other words, $P_{\mathbf{Z}}(\mathbf{Y}(t_2) \in B \mid \mathbf{Y}(t_1) \;\&\; \mathbf{X}(t_1)) = P_{\mathbf{Z}}(\mathbf{Y}(t_2) \in B \mid \mathbf{Y}(t_1))$, a.s., for every other system **X**.

This principle clearly entails C2, as the special case where an event in **Y** can be predicted with certainty, that is, its utd chance is one. Predictive locality merely generalizes C2 to cover all predictions, including those that are uncertain. The important theorem here is the following, although the converse also holds.

**6.1.5.1 Theorem** If all correlations w.r.t. chance are classical, then predictive locality holds.

*Proof*: If **X** and **Y** are correlated classically, then they are become independent upon conditioning on their joint state after they cease interacting. In that case, information about **X** is irrelevant to the later states of **Y**.■

In the next section we consider Bell's Theorem, which shows that if causal locality is true, then predictive locality is false.

### *6.1.6 Bell's Theorem*

One important consequence of predictive locality and causal locality together is that some measurement interactions are deterministic, including many that appear to be stochastic within the formalism of QM. Consider, for example, the EPR situation again, and suppose that we measure $A$ on **X** and $P$ on **Y**. From the outcome of either measurement the other can be predicted with certainty, even if the measurement events are far apart in space; but each individual measurement could, with non-zero chance, have a number of different results. The total state vector $|\Psi\rangle_{\mathbf{Z}}$ does not rule out any individual measurement results, but does rule out most *combinations* of measurement outcomes. This fact, together with predictive locality, entails that both measurement events are *locally* predetermined, as follows. The result of a *P*-measurement of **Y** can be predicted from the state of the world after *A* is measured on **X**, and so (if **Y** is isolated) this outcome can also be predicted from **Y**'s state before the measurement. But this is just to say that **Y**'s measurement has a locally determined outcome, and the same must hold for **X** as well.

One would not expect to be able to refute this consequence of predictive locality empirically, as it is usually impossible to distinguish between deterministic and stochastic systems experimentally. In particular, a system that appears to be stochastic may be merely chaotic, and since we cannot control the initial conditions with infinite precision it is practically impossible to tell the difference. It is therefore a remarkable achievement of John Bell to prove that, under the assumption of local determinism for experiments of this kind, we would expect different empirical results from those predicted by QM (Bell 1964).

Bell proved his theorem using a version of the EPR experiment devised by David Bohm (1951: 359).[10] A pair of spin-$\frac{1}{2}$ particles, **X** and **Y**, are prepared in a state where their spins are perfectly anti-correlated. This means that, if their spins are both measured (separately) with the same spin axis, the outcomes are guaranteed to be opposite. One will always be spin up, and the other spin down, regardless of the direction chosen. We imagine that the particles emerge from a central source in opposite directions, and then (when they are several metres apart) their spins are measured. We shall suppose that each measurement is on one of three different directions *A*, *B*, and *C*; the particles may be measured either in the same direction, or in different directions.

What does QM predict about this experiment? As stated above, when the spin measurements are parallel, such as when *A* is measured on both **X** and **Y**, the results should always be opposite. For measurements in different directions, we need to specify the directions *A*, *B*, and *C*. Let us suppose they all lie in the plane perpendicular to the motion of the particles, and at 120° to one another. If we measure *A* on **X** and *B* on **Y**, then QM gives a probability of $\frac{1}{2}(1 - \cos 120°) = \frac{3}{4}$ that the outcomes are the same, and indeed by symmetry *any* two measurements in different directions have the probability $\frac{3}{4}$ of yielding the same outcome.

What should we expect if each measurement is locally deterministic? In this case, each particle has what we may regard as a set of instructions, specifying the outcome it will produce for each measurement type *A*, *B*, and *C*. If we denote a spin-up result by '1', and a spin-down result by '0', then an instruction set will be something like ($A = 1$, $B = 1$, $C = 0$), or just (110) for short.

(The 'instruction' $A = 1$ is of course just the element of reality corresponding to the prediction that if $A$ is measured then the result will be spin-up.) It is clear that to account for the fact that measurements in the same direction always have opposite results, for **X** and **Y**, the instruction sets for **X** and **Y** must be opposite, that is, where **X** has a '0' **Y** has a '1', and so on. When the measurement directions are different, this deterministic hypothesis does not assign a probability to the measurement outcomes being equal, as it all depends on which instruction sets the particles have.

We can still get a frequency prediction from the hypothesis of determinism, however, as follows. Suppose we repeat the experiment many times, setting the measurement directions on each particle randomly and independently each time. For instance, on each trial we might roll a pair of fair, six-sided dice, one for each particle. Then on a 1 or a 2 we set the measuring device to $A$, on a 3 or 4 we set it to $B$, and so on. For this experiment, consider a particular instruction set (110) for **X**, so that **Y** has (001). For this set, there is a chance of $\frac{4}{9}$ that the measurement results will be equal, since there are nine possible pairs of measurements (each with the same chance $\frac{1}{9}$), including four on which the results will be the same. Moreover, by inspection, this chance of $\frac{4}{9}$ exists for every instruction set for **X** except (000) and (111), on which the chance is zero.

We have no information about the chance of each instruction set for **X**, but since the chance of the results being the same is no greater than $\frac{4}{9}$ on each set, this is an upper bound for the unconditional chance that the results are the same. According to QM, however, since the chance that the measurement directions are the same is $\frac{1}{3}$, the chance of equal outcomes is $0 + (\frac{3}{4} \cdot \frac{2}{3}) = \frac{1}{2}$. We thus obtain Bell's theorem, the surprising result that any locally deterministic theory actually yields empirical predictions that differ from those of QM. Moreover, real experiments similar to the one described here seem to show fairly conclusively that the true chance is much closer to $\frac{1}{2}$ than $\frac{4}{9}$.

If these experimental results are reliable, then the conjunction of causal with predictive locality must be false. If we continue to affirm causal locality, therefore, we must give up predictive locality. Of course, using Theorem 6.1.5.1 it then follows that the systems **X** and **Y** in the EPR experiment are non-classically correlated.

## 6.2 REACTIONS TO EPR

In section 6.3.3 I shall show that a firm commitment to local realism[11] does not require acceptance of predictive locality, or even C2, and thus a local-realist explanation of non-classical correlations will be given. First, however, we shall examine some other attempts to account for this phenomenon.

### *6.2.1 Bohr*

According to Bohr's colleague Leon Rosenfeld (1967: 142), the EPR paper 'came down upon us as a bolt from the blue.' Everything had to be dropped until this argument had been analysed, understood, and refuted. Within a few weeks Bohr had finished his reply, a paper with the same title as EPR's. As one might expect from the brief summary of the Copenhagen interpretation in §6.1.3, Bohr's attack was focused on the criterion of physical reality, which he alleged to contain an essential ambiguity. Where does this ambiguity lie? It is in the phrase 'without in any way *disturbing* the system.' Bohr agrees that, in the EPR experiment, there is indeed no 'mechanical' disturbance of the system **Y** after it has ceased to interact with **X**, so it cannot be mechanically disturbed by the event of **X** being measured. He holds nonetheless that **Y** is disturbed in *some* sense by the measurement of **X**, as this measurement is part of the total experimental arrangement that, according to principle (1) of the Copenhagen interpretation, is necessary to *define* physical quantities on **Y**. He notes, moreover, that physical quantities are used to predict the future behaviour of a system.

Bohr's paper is rather obscure and confusing, but his response does seem to be along the right lines, from my perspective. In denying any 'mechanical' disturbance of **Y** he clearly affirms causal locality, so his claim that **Y** is 'disturbed' in some other sense is likely to cause confusion. What other sense of 'disturbed' is there? Bohr regards **Y** as 'disturbed' when the values of its physical quantities are altered by the measurement on **X**. Now, according to principle (1) of the Copenhagen interpretation, the value of a quantity on **Y** is a *nonlocal* entity: it supervenes on the whole experimental apparatus, including system **X**, and is not confined to **Y**. Thus Bohr, while holding on to causal locality, seeks to deny some other kind of locality principle that bears at least a superficial resemblance to predictive locality.

Bohr's reply was not well understood by the physics community, and thus support for the Copenhagen interpretation was, in the long term, greatly weakened by this final exchange with Einstein. EPR's common-sense realist approach seemed far more wholesome than Bohr's. One physicist expresses his dissatisfaction as follows:

> This reply illustrates the strengths and weaknesses of the somewhat positivist approach of Bohr and his friends, with their emphasis on classical measuring apparatus. It enabled them to shrug off EPR, but at the cost, one might think, of refusing to face the issue. There is a way of proceeding in conceptual matters whose method is to define away any inconvenient difficulty. All the really tricky

questions are declared meaningless, despite the fact that they are sufficiently well comprehended to give rise to perplexity. On the EPR paradox it seemed that the Copenhagen school had achieved just such a Pyrrhic victory. (Polkinghorne 1984: 72)

This criticism is not entirely just, I believe, yet Polkinghorne's discomfort does seem representative of a large part of the physics community. Einstein's criterion of reality may have been wrong, as Bell's theorem and Aspect's experiment have shown, yet he did succeed in conveying his point that there is something very strange about quantum theory, something we do not yet understand. In short, I think Einstein's argument awakened the realist's desire for a deeper understanding of how these non-classical correlations can come about.

### *6.2.2 Bohm*

David Bohm (1952) formulated a theory that is similar to QM, but is nonetheless very much in the classical tradition. In his theory each particle has a well-defined trajectory, regardless of whether it is ever measured, and the motion is also deterministic. In spite of its deterministic nature, Bohm theory is not refuted by Bell's theorem, since the theory is non-local. Bohm theory seems to coincide with QM in its predictions regarding the EPR experiment – it is believed to predict the statistics obtained by Aspect, in other words. One might think that if one is prepared to sacrifice causal locality, then it is possible to explain Aspect's data without resort to non-classical correlations.

The basic idea of Bohm theory is rather simple. We retain the state vector, or wavefunction, from QM but it has a completely different interpretation. Instead of being a probability amplitude, it represents some sort of physical force field, analogous to the gravitational and electromagnetic fields. This $\psi$-field, as Bohm calls it, obeys the deterministic Schrödinger equation at all times – it never 'collapses.' In addition to the wavefunction, each particle also has a trajectory that is determined by a separate equation of motion. The velocity of the particle is directly dependent on the instantaneous value of the $\psi$-field, so that we might say that the particle is driven along by the field.

More precisely, the Schrödinger equation for a single particle in one dimension, $x$, is

$$i\hbar\frac{\partial\psi}{\partial t} = -\frac{\hbar^2}{2m}\frac{\partial^2\psi}{\partial x^2} + V(x)\psi.$$

Let us express the complex field $\psi$ in exponential form, that is, $\psi = R.e^{iS/\hbar}$, so that $S$ is the phase of the ψ-field. Then the equation of motion of the single particle is

$$\frac{dX}{dt} = \frac{1}{m}\frac{\partial S}{\partial x}\bigg|_{X(t)}.$$

We see then that the trajectory of the particle is determined by its initial position, together with the ψ-field. The particle is 'running on rails,' so to speak, which are built into the ψ-field.

It is well known that Bohm's theory violates causal locality – the non-locality has even been described as 'hideous'[12] – but it will be instructive to see exactly how this comes about. We must consider the dynamics of a pair of particles in Bohm theory, to see if forces applied to one can alter the motion of the other, even if they are far apart. (For simplicity, we assume that their masses are equal.) The pair of particles $X$, $Y$ will have a joint ψ-field in the two-dimensional configuration space of coordinates $(x, y)$, which obeys at all times the Schrödinger equation for a pair of particles, namely:

$$i\hbar\frac{\partial\psi}{\partial t} = -\frac{\hbar^2}{2m}\left(\frac{\partial^2}{\partial x^2} + \frac{\partial^2}{\partial y^2}\right)\psi + V(x, y)\psi.$$

Again writing the ψ-field in exponential form, the equation of motion of $X$ is now

$$\frac{dX}{dt} = \frac{1}{m}\frac{\partial S}{\partial x}\bigg|_{(X(t),\, Y(t))}.$$

The important detail to notice here is the point in configuration space at which the $x$-gradient of the phase field $S$ is evaluated. It is the point given by the instantaneous positions of not only $X$, but $Y$ as well! Thus, in some cases at least, the velocity of $X$ may depend functionally upon the position of $Y$, which is perhaps spatially quite distant from $X$.

Under what circumstances is there such a non-local dependence? First let us consider the case where the ψ-field $\psi(x, y)$ is the product of two separate ψ-fields, one for each particle. This is the case when the two particles have not interacted in the past. Then, if $\psi(x, y) = \psi_1(x)\psi_2(y)$, the phase $S$ of ψ will be a *sum* of phases, that is, $S(x, y) = S_1(x) + S_2(y)$. We then have that

$$\frac{\partial}{\partial x}S(x, y) = \frac{\partial}{\partial x}(S_1(x) + S_2(y))$$
$$= \frac{dS_1}{dx}.$$

Thus, in the case where $\psi$ is a product, the velocity of $X$ depends only on the gradient of the phase of its own $\psi$-field, evaluated at its own position, and so there is no non-local dependence on $Y$. When the two particles are correlated, however, in the sense that the $\psi$-field for the combined system $\langle X, Y\rangle$ does not factorize, there will be a non-local dependence of $X$ on $Y$ and vice versa.

What happens if we choose to apply some force to just one of the particles, such as $X$? Will the position of $Y$ be affected? We model forces applied to the particles, such as those due to magnetic fields for example, using the potential $V(x, y)$. If we apply a force to $X$ only, then $V$ will be a function of $x$ only. This potential has a causal influence over the $\psi$-field, according the Schrödinger equation, but in what way exactly? How will the introduction of a force on $X$ affect the $\psi$-field?

To investigate this, let us first assume that the particles have not interacted, so that the $\psi$-field is a product $\psi(x, y) = \psi_1(x)\psi_2(y)$. In this case, each particle has its own $\psi$-field and Schrödinger equation, and the Hamiltonian for $\psi_2$ is just

$$H_2 = -\frac{\hbar^2}{2m}\frac{\partial^2\psi}{\partial y^2},$$

that is, independent of $V$. Also, since in this case the velocity of $Y$ is determined by $S_2$ alone (as shown above), the trajectory of $Y$ is independent of $V$. Thus, when $\psi$ factorizes, the magnetic field potential $V$ has no effect on $Y$'s motion.

In the case where $\psi$ does not factorize, however, this argument cannot be used. Neither particle has its own $\psi$-field, and so the motions of both particles will be affected by $V$, even if this magnetic field exists in a part of space that is always far away from $Y$. We see that Bohm's theory is indeed grossly non-local, as a force applied to one particle can cause another particle, which may be arbitrarily distant, to change its motion.

It does not follow, however, that such a pair of particles may interact at a distance. The latter would involve a causal influence of $X$ on $Y$, or vice versa, which does not unambiguously happen here. It may instead be that the potential $V$ influences the $\psi$-field, which then alters the trajectories for both $X$ and $Y$. The correlation between the motions of $X$ and $Y$ may not be due to any causal interaction between them, but instead be due to their *common* cause $\psi$.

As noted above, in Bohm theory the $\psi$-field always obeys the Schrödinger equation, and thus does not 'collapse,' upon measurement, into an eigenstate of the quantity being measured. How does Bohm reconcile this with the fact that each measurement has a definite outcome? The rough answer is that the particle being measured in some sense 'selects' one of the eigenstates, and that the rest can subsequently be ignored, for all practical purposes; but first we must look at what quantum-mechanical properties in Bohm theory look like.

We have seen that the particle really has only one property, in itself, namely $\mathbf{x}$, its position (as well as its mass). What of all the other properties then, such as momentum, spin, and so on? We have already seen that the velocity of the particle is constrained to be equal to $(\nabla S(\mathbf{x}))/m$, so that we can consistently regard $\nabla S(\mathbf{x})$ as the momentum of the particle. Thus, the momentum of the particle is a property of the $\psi$-field in conjunction with the position of the particle (and, if it is correlated with other particles, with their positions too).[13] In a similar way, other properties such as spin also reside partially in the $\psi$-field.

If one measures the position of a particle, then one is simply measuring the quantity $\mathbf{x}$. What about measurement of other quantities, however, such as intrinsic spin? In these cases too, one is ultimately measuring $\mathbf{x}$, although under circumstances where $\mathbf{x}$ indicates the value of the other quantity. Spins, for example, are measured using the Stern-Gerlach apparatus, which involves deflecting a beam of particles in a magnetic field. The spin state of a particle is then determined by the direction in which it deflected by the magnet, either up or down, so that spins are measured ultimately by measuring positions.

Let us now consider the textbook treatment[14] of quantum measurement. Consider a system $\mathbf{X}$, whose state $\psi$ may be expanded using the eigenstates $|a_i\rangle_{\mathbf{X}}$ of $A$ as follows:

$$|\psi\rangle_{\mathbf{X}} = \sum_i c_i |a_i\rangle_{\mathbf{X}}.$$

We then suppose that a measurement of $A$ on $\mathbf{X}$ is performed by some measuring instrument $\mathbf{M}$, whose 'needle position' is represented by the operator $Q$. The eigenstates of $Q$ are $|q_0\rangle_{\mathbf{M}}$, $|q_1\rangle_{\mathbf{M}}$, $|q_2\rangle_{\mathbf{M}}$, etc. and we assume that the initial state of $\mathbf{M}$ is $|q_0\rangle_{\mathbf{M}}$, its 'zero' position. The initial state of the joint process $\langle\mathbf{X}, \mathbf{M}\rangle$ is therefore the product

$$|\Psi\rangle_{\langle\mathbf{X},\mathbf{M}\rangle} = \left(\sum_i c_i |a_i\rangle_{\mathbf{X}}\right) \otimes |q_0\rangle_{\mathbf{M}}.$$

If the measurement interaction between $\mathbf{X}$ and $\mathbf{M}$ is successful, then the final state of $\langle\mathbf{X}, \mathbf{M}\rangle$ will be

$$|\Psi\rangle_{\langle X,M\rangle} = \sum_i c_i |a_i\rangle_X |q_i\rangle_M,$$

so that the variables $A$ and $Q$ are correlated.

The problem is now for Bohm to explain how, from this final superposition, which never collapses, a definite measurement result emerges. The answer, of course, is that the position of the pointer $Q$ has a definite value at all times, since it is composed of a huge number of position-definite particles. Superpositions, for Bohm, exist only in the ψ-field – particles never suffer from such an indignity. Thus, each measurement always has a definite outcome, as the needle on a measurement apparatus always has a definite final position.

A more difficult problem for Bohm concerns the other elements of the superposition, which do not correspond to the particles' positions. In standard quantum mechanics these are assumed to vanish when the state vector is reduced, according to the Projection Postulate, but for Bohm they continue to exist for eternity. Will these parts of the ψ-field not lead to predictions that diverge from QM?

While Bohm admits that this could occur in principle, he maintains that, for all practical purposes, the other elements of the superposition may be assumed to vanish. To argue this he appeals to a feature of the QM formalism now known as *decoherence*. A measurement apparatus is a macroscopic object, and thus cannot be causally isolated from its environment. The interaction of a measuring instrument with the environment brings about a correlation between them, so that the measuring device loses its pure quantum state. We thus have to represent its state using a density matrix, as described in §6.1.2 (p. 154). Now, it is commonly argued that this density matrix rapidly evolves into a state where, expressed using a coordinate system of eigenfunctions of classical observables, it is diagonal. This means that, unless we also perform very detailed measurements of the environment, which is practically impossible, no measurement of the measuring device will show any interference between the components of the superposition. We are thus free to ignore, for all practical purposes, the elements that are not selected by the particle trajectories.

One aspect of Bohm theory we have not yet examined concerns the role of probability within the theory. Quantum mechanics, since it deals in physical chances, is able to predict (approximately and fallibly) the relative frequency of an outcome-type within a large set of repeated experiments. Clearly, if Bohm theory is to match this feat, then it must say something about probability. It does this via an additional postulate, which I call the *norm squared rule*. This rule says that, given knowledge of the ψ-field of a system at time $t$ and nothing else, one's personal probability density for the position of the particle is $|\psi(\mathbf{x}, t)|^2$.

The personal probability, of course, is the warranted degree of belief for a subject who has exactly this information. A postulate about warranted degrees of belief is a rather odd component of a physical theory, so let us look at how Bohm supports its inclusion. First, it should be pointed out (though Bohm does not mention this fact) that $|\psi(\mathbf{x}, t)|^2$ is a measure, so that when normalized it is (formally) a probability function. Second, this measure is invariant under Schrödinger evolution of the $\psi$-field, that is, the measure of any region of configuration space is equal to that of its image under the time-evolution operator. Thus, if the *initial* squared norm of the $\psi$-field is numerically equal to probability, then the same is automatically true for all later times as well. Bohm thus points out that the norm squared rule is self-consistent. Third, if one adds the norm squared rule to the other two postulates of Bohm theory, then one obtains a theory that is empirically equivalent to standard quantum mechanics.

We are now finally in a position to discuss how Bohm theory deals with the Bell inequality and the Aspect correlations. The situation here is rather simple. Since Bohm theory (it is claimed) is empirically equivalent to QM, it predicts that the Bell inequality will be violated in actual experiments, in agreement with the empirical results. Thus, it seems, these results do not force a renunciation of determinism. Instead we can give up both causal locality and predictive locality, neither of which holds in Bohm theory.

There are five difficulties with Bohm theory, which I will discuss in ascending order of severity. First, the equation of motion is a little odd in that the velocity, rather than the acceleration, of the particle is determined by the intensity of the field. This means that the momentum of the particle belongs mostly to the $\psi$-field – the particle has no momentum, or inertia, in itself. This is a stumbling block for some physicists,[15] who complain that this ruins the symmetry between position and momentum.

Second, as already mentioned, the theory explicitly violates causal locality. A measurement in one place can cause an instantaneous change in the velocity of another particle, which may be arbitrarily distant at the time. This is considered an undesirable aspect of any theory.

Third, Bohm theory as it stands is non-relativistic. A satisfactory theory would have to be Lorentz-invariant, like quantum field theory, but Bohm theory presents severe obstacles in this regard due to its non-local nature. This difficulty should not be regarded as fatal, however, at the present time, since a Bohm-style version of quantum field theory cannot be ruled out.[16]

Fourth, the norm squared rule is not legitimate as part of a physical theory. A physical theory should describe the physical world, and not include (logically separate) normative claims about what humans should believe. If the norm squared rule could be derived from the other two postulates then it would be acceptable, but this is not the case. It should be noted here that the norm squared rule is similar to Miller's principle, which I use to explain relative frequencies, but there is an important difference. Miller's principle follows logically from the definition of chance, whereas the norm squared rule cannot be derived from the definition of the $\psi$-field.

Fifth, the theory does not *explain* the observed data, even if we include the norm squared rule. To explain a phenomenon *E*, we should recall from chapter 1, is to infer *E* from a maximal description of *E*'s causes. Now the causes of an event, according to Bohm theory, are the initial state of the $\psi$-field, the initial positions of the particles, and the dynamics of the field and the particles. The predictions of relative frequencies, however, are not based on maximal knowledge of these facts. Instead the predictions depend upon the right kind of ignorance, namely, ignorance of the particle positions, together with the norm squared rule (which has nothing to do with causation). Bohm theory does provide rules for inferring relative frequencies, but does not thereby *explain* those frequencies.

It may seem odd that physicists have not been as critical of Bohm's norm squared rule as philosophers have been of Miller's principle. The reason for this, I believe, is that Bohm's rule is merely a continuation of the long-established tradition of such rules in statistical mechanics. It is unfortunate that, at the time when it became necessary to introduce probabilities into physics, determinism was so firmly entrenched. This made the reliance on assumptions about epistemic probabilities inevitable, though hardly good physics.

Bohm theory is almost certainly false, for the reasons noted above, yet it is very instructive. For one thing it shows just how hard it is to account for the EPR correlations within the paradigm of classical physics, with complete models of well-defined trajectories. In particular, it shows that such a theory has to violate predictive locality, and also to smuggle in assumptions about relative frequency.

### *6.2.3 Everett*

Like Bohm, Everett (1957) claims to have found a theoretical framework in which the EPR correlations make sense. The central part of his project, which is extremely ingenious as well as ambitious, is a reformulation of quantum mechanics itself, which he calls the *relative state* formulation. Everett is critical of the conventional formulation of quantum mechanics, which he calls the 'external observation' formulation, on account of its silence regarding closed systems. He claims that '[t]he whole interpretive scheme of that formalism rests upon on the notion of external observation' (1957: 316), and this criticism seems to be just. We have seen that the Copenhagen interpretation is quite explicit about this, with its view that the properties of quantum systems depend essentially upon the macroscopic experimental arrangement, and particularly the measuring instruments with which they interact. Everett attempts to construct a universal quantum mechanics, the foundations of which are quite free of the concept of observation, and this is surely a noble task.

The relative state formulation of quantum mechanics is remarkably simple. It is actually just quantum mechanics, without the Projection Postulate. A measurement is treated in exactly the same way as any other physical interaction, as a physical process governed entirely by the Schrödinger equation. Now, since in the standard formalism a system evolves stochastically only when a measurement is made it follows that, in the relative state formulation, all systems are deterministic. Thus, in one respect Everett's theory is similar to Bohm's, but there are also very large differences, as we shall see.

A second similarity between Everett and Bohm is that Everett also interprets the wavefunction as an objective, physical entity. It does not, in any sense, represent an epistemic state. Everett has no separate particles in his ontology, however; the $\psi$-field is all there is.

The main difference between Everett's theory and Bohm's is that Bohm does not allow a particle's state to be a superposition. According to Bohm, although a particle is driven along by the wavefunction in a manner that is unpredictable for us, it still has a well-defined position at all times. Knowledge of the wavefunction may not enable us to predict exactly where the particle will be measured, so that there may be some epistemic probability of finding it 'here,' and some probability of finding it 'there,' but its position cannot be a superposition of 'here' and 'there.' On Everett's view, by contrast, such genuine superpositions are very common.

A second difference between Everett and Bohm is that Everett's theory satisfies causal and predictive locality. The Schrödinger equation is the only dynamical law, according to Everett, and it is fully local. Of course there is a *kind* of non-locality about the $\psi$-field, as two systems that previously interacted will share a wavefunction that does not factorize, even if they are now spatially quite distant. This should be called something else, however, such as 'nonseparability,' as it is quite different from either causal or predictive locality.

The really novel feature of Everett's approach is his explanation of how the classical world, with its precise values for macroscopic quantities, emerges from the $\psi$-field. This is a serious issue for any view which maintains that quantum mechanics is to be applied to every system without exception, even measuring instruments. Since, after measurement, the 'needle' on the measuring instrument will have no definite position, according to the ψ-field, how can we reconcile this with the common-sense reality in which it does?

Everett's answer to this question is as follows. First he points out that common-sense reality, that is, the classical world, is fundamentally a matter of the way things appear to *us*. We look at the needle and *see* that it has a definite location, rather than being smeared out perhaps. He then considers what happens, according to his theory, when an observer looks at the measuring instrument at the end of the experiment. According to the Schrödinger equation, the observer is also then 'sucked into' the superposition! Assuming, for simplicity, that the system measured has been absorbed into the measurement apparatus, we get a state for the measuring device + observer as follows:

$$|\Psi\rangle = \sum_i c_i |m_i\rangle |o_i\rangle,$$

where the states $|m_i\rangle$ of the apparatus are ones like 'the needle has position $q_i$,' and the states $|o_i\rangle$ of the observer are ones like 'I see the needle in position $q_i$.'

Everett then (having defined his relative state as in §6.1.2) points out that, although neither the apparatus nor the observer has a definite, common-sense state, each does have such a state *relative to the other*. More precisely, for each state $|o_i\rangle$ of the observer there is a well-defined relative state $|m_i\rangle$ of the apparatus. What is it like to be such an observer? What is the nature of his conscious experience – will he see a smeared-out needle? According to Everett, each vector $|o_i\rangle$ of the observer is associated with a separate locus of consciousness, each of which sees a definite position for the needle. Thus, for instance, the 'mind' associated with $|o_i\rangle$ sees the needle in position $q_i$, where $q_i$ is the eigenvalue corresponding to $|m_i\rangle$.

When the observer looks at the measuring instrument, then, his ψ-field branches as it is entangled with the ψ-field of the instrument. He does not notice this, however, as each branch of the wavefunction is associated with a separate mind, each of which is unaware of the others (at least before 1957!).[17]

Everett does not discuss the problem that, even after the branches of the superposition become 'macroscopically distinguishable,' they may still be liable to interfere with one another, giving rise to predictions different from QM. I assume however that, like Bohm, he is relying on the decoherence of the measuring instrument's density matrix. This, for all practical purposes, makes it impossible to observe any macroscopic quantum phenomena, even though they are not ruled out in principle.[18]

The 'classical world,' or rather worlds, that emerge in this way from the universal ψ-field are not entirely in accordance with common sense. First, the future is entirely indeterminate. This does not mean merely that the future is unpredictable from the past, but something much more shocking: there simply are no facts about what will happen. If I assert, in March 1997, that the Labour Party will form the next government, then this is likely neither true nor false, since the Labour Party wins the next election in some future branches of 'me' but not others. We are familiar with the notion that such assertions about the future may be impossible to verify, until the time comes of course, but the idea that they lack truth values is counter-intuitive.

A more serious oddity of Everett's worlds is that the past is also, albeit to a lesser extent, indeterminate.[19] Since branches may converge as well as split, there are alternative pasts as well as alternative futures. In general, we might say that Everett's view is rather solipsistic, in a sense. The only shared, objective reality is the $\psi$-field, which is completely unfamiliar to us. The common-sense world turns out to be merely *my* common-sense world, having little reality beyond my experience of it. The appearance of a shared, familiar world is an illusion.

Perhaps the greatest virtue of Everett's theory is that it accounts for the EPR correlations without invoking any non-local causal interactions. Given that the formalism satisfies both predictive and causal locality, one may wonder how this is possible – does Bell's theorem not rule out such theories? The point to realize is that it is the evolution of the $\psi$-field that satisfies causal and predictive locality, not the classical world; and the EPR correlations exist in the classical world. Everett's classical world evolves neither deterministically nor stochastically – these categories simply fail to apply. The future of the classical world cannot be determined by the past, as the future is not even determinate. Yet, for the same reason, the classical world does not evolve stochastically either. Instead, the classical world branches into a large number of distinct futures, all of which are equally real as they are all part of the $\psi$-field. It is not a matter of *chance* which of these futures 'I' end up experiencing; rather, each is experienced by someone, although none of these awarenesses can really be identified with me since identity is a transitive relation.

If we examine the EPR experiment from Everett's point of view, then we find that, in each classical world, spins measured in the same direction give opposite results. It does not follow from this, however, that each spin measurement had a determined outcome. Rather, each spin measurement has both possible outcomes, as upon measurement each apparatus becomes correlated with the particles, and enters the joint superposition. The important point is that, *relative* to the state of having recorded spin UP on one device, the other device has the unambiguous state spin DOWN, and vice versa. Thus, in any classical world, as defined by Everett, the instruments display opposite readings.

The analysis of spin measurements in different directions is more tricky, as in this case quantum mechanics gives a chance $\frac{3}{4}$ that the outcomes will be equal, which supports a prediction that the long-run relative frequency of equal results will be about $\frac{3}{4}$ as well. Since the classical world evolves neither deterministically nor stochastically, on Everett's view, how does he predict such relative frequencies?

It is clear that, in Everett's picture, if the EPR experiment is repeated many times, using the measurement directions *A* and *B* say, a frequency for equal outcomes of about $\frac{3}{4}$ will exist in some final branches of the superposition but not others. Indeed, for every sequence of outcomes that we generally regard as possible, there will be a branch containing exactly that sequence. How then can Everett claim that his theory predicts, or explains, the datum $\frac{3}{4}$ that we actually observe? Could it not, just as easily, be any other value from 0 to 1? In response to this problem Everett says:

> In order to establish quantitative results, we must put some sort of measure (weighting) on the elements of a final superposition. This is necessary to be able to make assertions which hold for almost all of the observer states described by elements of a superposition. We wish to make quantitative statements about the relative frequencies of the different possible results of observation – which are recorded in the memory – for a typical observer state; but to accomplish this we must have a method for selecting a typical element from a superposition of orthogonal states. (1957: 321)

The rough idea, then, is to have a probability measure over the final branches, or classical worlds, such that the collection of branches with frequencies close to $\frac{3}{4}$ gets most of the weight. For a very long sequence of experiments, we might have, for example, that the measure of the class with frequencies in [0.74, 0.76] is 0.99. In this way Everett can assert that the statistics one actually observes exist in 'almost all' worlds, where 'almost all' refers to the measure rather than the cardinal number.[20]

Which measure does Everett take? Like Bohm, he considers that any viable candidate must be invariant under Schrödinger evolution, which of course leaves us with only the familiar squared norm of the $\psi$-field. (Let us call this 'Born measure,' for convenience.) This measure has the added bonus that it yields frequency predictions identical to those of QM, so Everett's theory predicts that the Bell inequality will be violated (as it is violated in 'almost all' worlds).

The reader may notice some similarity between Everett's and Bohm's use of the Born measure to predict relative frequencies. Indeed Everett, like Bohm, refers to the use of Lebesgue measure in statistical mechanics by way of justification: 'Our procedure is therefore quite as justified as that of classical statistical mechanics' (1957: 322). Now the comparison with statistical mechanics is appropriate for Bohm, since his theory is deterministic and directly parallel to statistical mechanics. The probabilities he introduces are epistemic, and arise due to our ignorance of the exact initial conditions. Although his identification of epistemic probability with Born measure is unjustified and illegitimate, it is at least meaningful. Everett, by contrast, cannot match this level of success, as we shall now see.

What would an identification of epistemic probability with Born measure *mean*, within Everett's theory? Ideally it would work in the same way as in Bohm theory, specifying warranted degrees of belief about future experimental outcomes given knowledge of the current wavefunction, for that is the usual manner in which quantum-mechanical probabilities are used to predict frequencies. In this case probabilities would be defined for statements such as 'If I repeat the experiment 10,000 times, then the relative frequency of $a$ will be $\frac{3}{4}$.' The problem here, however, is that such statements are actually meaningless in Everett's picture, as there is no unique future, and so such statements lack truth values. If Everett were to identify Born measure with epistemic probability, therefore, this move would be not merely unjustified but actually nonsensical.

To his credit, Everett does not attempt such an identification, and leaves the Born measure uninterpreted. This being so, however, it is quite unclear how the measure is used to predict relative frequencies. An inference from 'this outcome has Born measure 0.99' to 'this outcome will occur' is simply invalid, unless a suitable interpretation for Born measure is provided. This inference problem for Everett looks very severe, particularly when we bear in mind that the only known way to infer frequencies from objective probabilities is via Miller's principle, an approach that is unavailable to Everett.

Everett's theory has been criticized on other grounds, such as its ontological profligacy, its revisionary ideas about reality and actuality, and its need for a preferred orthonormal basis for an observer system, but the problem of predicting relative frequencies seems to be the most devastating. As it stands, at least, the theory does not save the phenomena.

### *6.2.4 Other Approaches*

The three responses to the EPR argument discussed above are all several decades old, at least, although each still has able contemporary supporters. I consider them to be the best work on the subject to date, but for the sake of completeness I will briefly mention one other, more recent type of response.

The main idea of this final approach is to regard the wavefunction as a real field, as do Bohm and Everett, but hold that the Schrödinger equation is of limited validity. Thus, the Schrödinger equation is obeyed, for all practical purposes, in the atomic and subatomic world, but at some larger scale is superseded by another rule. This other rule prescribes a stochastic localization of the wavefunction, in some manner that at least approximates the von Neumann collapse. Thus, roughly speaking, the state vector reduction is seen as a physical, rather than epistemological, process. One view of this kind has been advanced by Ghirardi, Rimini, and Weber (1986), and another is due to Roger Penrose (1994: chap. 6).

## 6.3 BEYOND POSTULATE CSM3

We have seen that nature suffers non-classical correlations between systems, whereas CSM does not. It follows that CSM is false, although I believe it is approximately true within a restricted domain of 'large' systems. If CSM is false, it follows that at least one of the postulates on which it is based is false, and I think that postulate is CSM3. This postulate allows a history of a composite process always to be factorized into histories of the subsystems, which is a necessary assumption in the proofs of the independence theorems 5.4.3 and 6.1.1.2. This section has two aims, therefore. It must be shown that (1) it is possible to reject CSM3, and (2) if CSM3 is false, then non-classical correlations are possible. The rejection of CSM3 will then be a major component of the interpretation of QM presented in the next chapter.

### *6.3.1 Factorizability and Consistent Families*

I have already argued, in §3.3.7, that the intuitions leading us to accept CSM3 are flawed. If two systems interact then they are a single dynamical system, so the actual history that emerges will be a history for the pair of systems. This joint history may factorize, but there is no guarantee that it will. The intuition that it must factorize depends upon a conflation of concrete objects with states of affairs, or upon the erroneous idea that states of affairs are complete.

These points are sufficient, I believe, to show that it is feasible to reject CSM3. Then, since (as shown in the next two sections) this rejection allows us to explain non-classical correlations, we have very strong grounds to do so. One may yet feel unsatisfied, however, with this explanation of quantum correlation, since one still lacks any real *insight* into its cause. While the paradox has been removed, we do not really see what is going on. Unfortunately, it is part of the view itself that we will *never* understand quantum correlation in the way we understand a classical dynamical system, such as a steam engine. The correlation arises from the non-factorization of the joint history, which entails that no understanding of how the systems are in themselves is possible. In the remainder of this section I shall, however, try to provide as much insight into the situation as I can.

It may help, for example, to look at a non-factorizable proposition concerning a pair of systems, such as the following. I make two mugs of coffee and absent-mindedly put a lump of sugar into one of them. A few moments later I remember adding the sugar, but am unsure which mug it went into (they both look the same). If the mugs are **X** and **Y**, then I know that $\mathbf{Z} = \langle \mathbf{X}, \mathbf{Y} \rangle$ contains one sugar cube, but I do not know the number for **X** or for **Y**. My knowledge can be represented as $\mathbf{X} + \mathbf{Y} = 1$ or, alternatively, as $(\mathbf{X} = 1 \ \& \ \mathbf{Y} = 0) \vee (\mathbf{X} = 0 \ \& \ \mathbf{Y} = 1)$. This does not factorize into knowledge about **X** and knowledge about **Y**.

Examples of this kind, while showing that knowledge of a composite system may not factorize, do not help us very much. For one thing the knowledge is clearly sub-maximal, as it may be expressed in the form of a disjunction, so that it is entailed by each disjunct. States and histories, on the other hand, are maximal propositions. It will be noted that each disjunct is factorizable, so one might even suspect that the failure to factorize is due to the lack of maximality. Also the 'correlated' systems – the mugs – need not have interacted in any way and bear no special physical relation to each other. The correlation in such examples is purely an artefact of the way we have gathered information about the systems, and does not depend upon the systems themselves. In the case of non-classical chance correlations, there does seem to be a physical relation between the correlated systems, due to their previous interaction.

To get some insight into why even a maximal description of a system $\langle \mathbf{X}, \mathbf{Y} \rangle$ may fail to factorize let us consider the fact that physical quantities seem to represent relations rather than intrinsic properties. Let us examine length, as it is perhaps the simplest example of a physical quantity. What is a length? It is a number, perhaps a quotient or a real number, that attaches to a body such as a straight rod. How does it become attached, however? A rather simplistic version of the story is as follows.

Given *two* straight rods, such as **X** and **Y**, we can define the *relative length* $R(\mathbf{X}, \mathbf{Y})$ as follows. Take a rod $\mathbf{Y}'$ that is similar to **Y**, in the sense that when placed side-by-side with **Y** both ends of $\mathbf{Y}'$ can touch an end of **Y** at the same time. Place **Y** and $\mathbf{Y}'$ end-to-end so that they form another straight rod. The rod $\mathbf{Y} - \mathbf{Y}'$ can be compared with **X** to see if it is longer or shorter, as follows: Place **X** and $\mathbf{Y} - \mathbf{Y}'$ side-by-side so that one end of **X** touches one end of $\mathbf{Y} - \mathbf{Y}'$. If the other end of $\mathbf{Y} - \mathbf{Y}'$ touches an interior part of **X** then **X** is the longer rod. If the other end of **Y** − **Y′** touches no part of **X** then **X** is shorter. If it touches the other end of **X** then **Y** − **Y′** and **X** are similar. If **Y** − **Y′** is shorter than **X**, then get another rod **Y″**, also similar to **Y**, and form the rod **Y** − **Y′** − **Y″**. Continue this process until the two rods are about the same length. *R*(**X**, **Y**) is then the number of **Y**-rods that together are similar to **X**.

Clearly, for some pairs of rods the number *R*(**X**, **Y**) will not be very precise, whereas for others it will be quite accurate. Let us just consider those pairs for which is it fairly precise, in the sense that **X** is, as far as we can tell, similar to some concatenation of **Y**-rods. We can then extend the relation R to all pairs of rods by stipulating the following:

$$R(\mathbf{X}, \mathbf{Y}) = \frac{1}{R(\mathbf{Y}, \mathbf{X})}, \text{ and } R(\mathbf{X}, \mathbf{Y}) = \frac{R(\mathbf{Y}, \mathbf{W})}{R(\mathbf{Y}, \mathbf{W})}.$$

Thus, if previously *R*(**X**, **Y**) was defined, *R*(**Y**, **X**) is now also defined. If previously *R*(**X**, **Y**) was not defined, then find some (perhaps very short) **W** such that *R*(**X**, **W**) and *R*(**Y**, **W**) are both defined, and then define *R*(**X**, **Y**) as the ratio of these.

Now, in making stipulations such as these there is no a priori guarantee that they will even be consistent. For instance, there might be two non-similar rods **W** and **W′** such that *R*(**X**, **W**), *R*(**Y**, **W**), *R*(**X**, **W′**), and *R*(**Y**, **W′**) are all defined, but *R*(**X**, **W**)/*R*(**Y**, **W**) is different from *R*(**X**, **W′**)/*R*(**Y**, **W′**). As a matter of *empirical fact*, however, these stipulations do turn out to be consistent, within the limits of experimental error. Since the two relations above do hold for the family of relations *R*, I shall say that *R* is a *consistent family*.

The fact that *R* forms a consistent family allows us to define length as a property, attaching to a single body, as follows. We select some rod – it does not matter which one – and designate it as the unit. We may, for instance, choose the popular rod called 'metre,' or **M** for short. The length of a rod **X**, *l*(**X**) say, is then defined to be *R*(**X**, **M**). It must be stressed that the unit for length, like all units, is a concrete body and not an abstract number.

The fact that *R* forms a consistent family also allows us to factorize statements involving length about composite systems. The statement *R*(**X**, **Y**) = 2.64, for instance, might be expressed as the conjunction *l*(**X**) = 12.91**M** & *l*(**Y**) = 4.89**M**.

The example of length is unusually simple (although it is still a good deal more complex than as portrayed here), yet the essential features are apparently shared with all physical quantities (most obviously for quantities with units). The factorizability of states of affairs involving such relations, therefore, depends upon their forming a consistent family.

Why do relative lengths form a consistent family? A common-sense realist answer to this question might be as follows. The relative length *R*(**X**, **Y**), as defined above, is caused by the individual properties of the bodies **X** and **Y**. If, for instance, **X** is three times as long as **Y**, then this relation is brought about by the *individual* concrete natures of the two bodies. Moreover, it seems that all rigid, straight rods are commensurable with respect to the relations of relative length; that is, any two rods have a well-defined relative length. Since then the relation *R* is grounded in the objective natures of individual bodies, which are all pairwise commensurable, we should expect the kind of stability that a consistent family of relations has. This is exactly the right kind of 'stability' that enables each rod to be assigned its *own* length, relative to some unit. We should expect relative lengths to reduce to these quasi-absolute lengths, since relative lengths are ontologically grounded in the individual natures of the bodies.

This answer, though rough, is basically correct in my view. It should be noted that the 'concrete natures' of the individual bodies, which determine the relative lengths, cannot themselves be represented by numbers. It simply does not mean anything to assert that the length of this rod is the (dimensionless) number 17.62. We thus have the paradox that, while lengths are ultimately grounded in the *individual* natures of bodies, their numerical representation is irreducibly *relational*. This paradox is easily resolved, however, provided we recognize the distinction between concrete reality and abstract states of affairs. We must resist the temptation to think of physical quantities, that is, numbers, as existing in concrete systems. Numerical lengths, which are mere representations, are irreducibly relational, whereas the concrete natures, or 'real lengths,' belong to individual bodies.

If the 'real length' of a body, which is part of its concrete nature, cannot be represented by a number, then how is it to be understood? It is hard to see how it could be represented in abstract terms, even in principle. The best, most versatile, models we have are mathematical structures, and yet these are only able to model relations between concrete systems. The individual natures that underlie these relations seem to be forever hidden from our view. (We note therefore that mathematical models cannot be complete representations of concrete systems, even in principle.)

We have seen that a description of a system ⟨**X**, **Y**⟩ involving physical quantities will factorize provided the relations in question form a consistent family. On the other hand, if a maximal description of ⟨**X**, **Y**⟩ requires mention of some relation that is not part of a consistent family, then the description will not factorize. Which relations might not be part of a consistent family, however? Do such relations exist? I think the following example is useful here.

Suppose a sheet of paper is torn roughly into two pieces, **X** and **Y**.[21] There is then a similarity between **X** and **Y** that is of such a particular and idiosyncratic kind that probably no other object in the cosmos is similar to them in this way. (Unless you perhaps take a photograph of one of them, or some such thing.) This relation is therefore not part of a consistent family and cannot be directly factorized.

This relation may be factorized, of course, if we consider **Z** = ⟨**X**, **Y**⟩ as a collection of fibres. (**Z** is then viewed as a very large collection of systems, rather than a mere pair.) Now the fibres are described adequately by their relative lengths, breadths, positions, and so on, all of which form consistent families. The description of **Z** using such quantities is therefore fully factorizable, and so a description of **Z** at the level of fibres will factorize into descriptions of **X** and **Y**. The latter two propositions will jointly entail, of course, that **X** and **Y** fit together exactly.

If we are being pedantic, it should be noted that the above argument assumes that each torn fibre breaks cleanly, without any irregular, jagged edge, as otherwise the shift of focus from the sheets to the fibres does not help. What if, however, the fibres do tear in an irregular way, so that the two halves of a single torn fibre bear a unique relation to each other? In this case the focus will have to be shifted again to an even smaller scale where, with any luck, such awkward *sui generis* relations do not exist.

Will such a level of description be found, where the relations between the systems all belong to consistent families? For the torn sheet of paper I think that such a level of description does exist, as least for all practical purposes. Thus, a maximal description of the two jagged pieces will factorize. I see no reason to hope that this will be true in general, however, for all pairs of systems that have interacted with each other. In particular, if the systems **X** and **Y** are small (perhaps they are photons), then there is far less scope for finding other descriptions at smaller scales. If, as a result of their common origin, a pair of photons can have a *sui generis* relation like that between the two halves of the torn sheet of paper, then it is likely that their maximal description will not factorize.

As in the case of length, my view is that the *sui generis* relation between the photons is grounded in the concrete nature of each photon. However, due to the way mathematical models relate to real systems, via relations between systems, these individual natures cannot be represented individually.

The above analysis of relational measures is admittedly crude, but I think it offers some insight into quantum correlation. It leads us to expect that CSM3 will break down for small systems that have interacted in the past. What is meant by 'small' here? It is not a matter of physical dimension, as in that case a pair of photons ten metres apart might be considered a large system! It is rather a matter of composition; that is, if **X** is a subsystem of **Z**, then **X** is smaller than **Z**. The distinction between classical and quantum systems is discussed in more detail in section 7.2.1.

### *6.3.2 Factorizability and Predictive Locality*

The reason why CSM3 is so important to the issue of non-classical correlation is that if factorizability fails then so does predictive locality. This is most easily shown in the coffee mug example. By Theorem 6.1.5.1 it then follows that non-classical correlations are possible.

Suppose each mug is measured for sugar content, perhaps by someone tasting it. What will I predict for each measurement outcome, if my state of knowledge $K$ is given by the proposition $\mathbf{X} + \mathbf{Y} = 1$? In view of the symmetry of the mugs, it seems reasonable to put $P_K(\mathbf{X} = 1) = P_K(\mathbf{Y} = 1) = \frac{1}{2}$. What if I learn of the result of **X**'s measurement before that of **Y**, however? If I learn that $\mathbf{X} = 1$, for instance, then I can infer with certainty that $\mathbf{Y} = 0$, and if I hear that $\mathbf{X} = 0$ then I infer that $\mathbf{Y} = 1$. Within the epistemic state $K$, in other words, information about the system **X** is highly relevant to the predictions I make about **Y**, which is just to say that predictive locality fails for $K$.

Now, the state $K$ is not maximal, of course, but this is not relevant to the issue here. The relevant fact is that $K$ does not factorize. This means that the information in $K$ is non-local, so that the predictions supported by $K$ are also non-local. This non-locality of information does not depend upon any causal non-locality, of course, as these statistics for $P_K$ obviously do not require any mysterious causal interaction between the mugs while they are being tasted.

### *6.3.3 Predictive Locality and Local Realism*

There is a consensus that Bell's theorem, together with the Aspect results, rules out a view that d'Espagnat (1979: 158) calls *local realism.*[22] Local realism consists of three separate theses, which d'Espagnat calls *locality, realism,* and *induction*. By 'locality,' he means causal locality, the claim that no causal influence can propagate faster than light. d'Espagnat's definition of realism is that 'regularities in observed phenomena are caused by some physical reality whose existence is independent of human observers' (1979: 158). In the case of the EPR correlations, which count as a 'regularity,' this surely means that they are due to some common cause. They must have some cause or other, according to realism, and locality rules out any direct interaction between the two systems, so that we are left with a common cause. (Induction is just the view that the physical chances of event types can be estimated by measuring relative frequencies of those event types in repeated trials.)

The argument against local realism given by Bell and d'Espagnat assumes that realism includes the EPR criterion of physical reality, discussed in §6.1.4. As we have seen, the application of this criterion to the EPR experiment yields the conclusion that each spin measurement is locally predetermined. The general layout of the argument may therefore be represented as follows:

| | | |
|---|---|---|
| 1 | local realism $\Rightarrow$ local determinism | Using the EPR reality criterion |
| 2 | local determinism $\Rightarrow$ the Bell inequality | Bell's theorem |
| 3 | Not(the Bell inequality) | Aspect experiments |
| $\therefore$4 | Not(local realism) | From 1,2,3. |

I have shown, however, that the EPR reality criterion is dependent upon the principle of predictive locality, so that where predictive locality fails it does not provide even a sufficient criterion of physical reality. Predictive locality, in turn, depends upon CSM3, since it fails if histories of composite processes do not factorize. Finally, we have seen that CSM3 is probably false in general, even if realism is true. Thus, realism does not entail predictive locality, and so local realism is not refuted by Bell's theorem and Aspect's results. This may be summarized as follows:

| | | |
|---|---|---|
| 1 | (local realism & predictive locality)<br>$\Rightarrow$ local determinism | |
| 2 | local determinism $\Rightarrow$ the Bell inequality | Bell's theorem |
| 3 | Not(the Bell inequality) | Aspect experiments |
| $\therefore$4 | Not(local realism & predictive locality) | From 1,2,3. |

# 7

# The State Vector

In the previous chapter on correlation we saw that the chance function in quantum mechanics does not always satisfy predictive locality. I offered an explanation for why this might be so, but did not give any interpretation of the quantum-mechanical formalism. In this chapter I will develop a theory of what the quantum state vector represents. I shall look at the two main views on this, that (1) it models some physical field, as Bohm, Everett, and R. Penrose maintain, and that (2) it represents 'knowledge,' or 'information' about the system. My view, perhaps surprisingly, is that both (1) and (2) are correct.

As is well known, many of the conceptual difficulties of quantum mechanics are centred around the issue of measurement. This cluster of puzzles is often called the *measurement problem* of QM. The measurement problem is impossible to separate from the question of how the state vector is to be interpreted, so I will have to say something about measurement as well.

Among the standard range of interpretations of QM, my own view is closest to the Copenhagen interpretation. I agree, at least roughly, with the five principles laid out in section 6.1.3. Nonetheless, I reject the instrumentalist and positivist aspects of Bohr's philosophy. There is far more going on in the world than can be represented in mathematical models.

## 7.1 THE PROBLEM

The state vector represents, apparently, some sort of physical field, as will be argued in the next section. Also, however, it is closely connected to epistemic probability, so that it represents the possible knowledge that one can have about the system, as I argue in §7.1.2. These two facets of the wavefunction seem to be in contradiction, forcing us to choose between them. Physical facts and epistemic states are quite different things. The latter makes an essential reference to minds, since knowledge requires a knower, whereas one would usually say that a physical fact is independent of mind. I will show in §7.3.4, however, that the causal theory of chance allows them both to be correct.

### *7.1.1 The Physical Interpretation*

I shall advance two arguments that the wavefunction represents a real, physical field of some kind.

First, the wavefunction has a deterministic equation of motion, which has been used to predict a wide range of phenomena to a very high degree of accuracy. Consider, for instance, one of the early successes of wave mechanics, the explanation of Balmer's formula for the spectral lines of hydrogen. These energy differences fall right out of the Schrödinger equation, using an appropriate static potential well. Moreover, the individual solutions to the time-independent wave equation (which correspond to the electronic energy levels) make physical sense, being similar to standing waves. Since the wave is confined to the well, and must be continuous in $\psi$ and $\nabla\psi$, only certain discrete frequencies (energies) are possible, which correspond exactly to those observed. It should be noted that this 'prediction'[1] does not involve probabilities, and indeed was made prior to Born's statistical interpretation of $\psi$.

Schrödinger, of course, did not 'cook up' his equation to generate these data, but was working with a very general, theoretical heuristic, based on the assumption that the classical Hamilton-Jacobi equation gives an incomplete and approximate description of $\psi$. He took that equation to describe only the phase of the wavefunction, and also imposed a condition of linearity, which led to his own equation. The 'prediction' of the Balmer series was a remarkable success for such a deep, theoretically motivated equation, and it is but one of a galaxy of such triumphs. If there is ever evidence of correspondence between a model and reality, this must be such a case. It simply will not do to say that the standing waves within a hydrogen atom are 'waves of knowledge,' since these waves are causes of physical phenomena, and one's knowledge of a system does not have any causal power.

For the second argument, we grant the fact that the wavefunction can be used to calculate probabilities of measurement outcomes. It does not follow from this, however, that the wavefunction is just a means for summarizing what we know about a system. The essential point here is that the wavefunction is *complex*-valued, and so involves what are called 'probability amplitudes' rather than probabilities. The use of amplitudes rather than probabilities, moreover, is not a mere mathematical convenience but is necessary for empirical adequacy.

An analogy with geometry may be helpful here. The basic quantity in geometry is the distance between a pair of points. A geometrical representation of a rigid system of particles, however, is not usually a list of such distances, one for each pair of particles in the system. (Let us call this the *d*-list.) Instead we give a set of coordinates $(x, y, z)$ for each particle, from which the *d*-list can be calculated using the Euclidean metric, as the square root of the sum of the squares of the coordinate differences. Now what are we to say about these coordinates? Are they merely a convenient way of summarizing all our information about the distances? It seems that the coordinates are rather more significant than this, that they provide a *deeper* representation[2] of the system than does the *d*-list, for the following two reasons.

First, the coordinates manage to generate the *d*-list using fewer numbers. If there are $N$ particles in the system, then there are $\frac{1}{2}.N(N-1)$ distances, but only $3N$ coordinates. For any $N$ greater than 7 the coordinate description is shorter, and the difference becomes very significant for large $N$. With $N = 50$, for example, the *d*-list contains 1225 numbers, whereas there are only 150 coordinates. It seems that the coordinate description, which includes the Euclidean metric, has some built-in knowledge about the structure of space. The coordinate description 'knows,' for example, that space is three-dimensional and Euclidean.

Second, the coordinate structure can be used to make predictions of distances that have not been measured, as follows. A coordinate system is determined by assigning coordinates to any four non-coplanar points, which we may call $O$, $X$, $Y$, and $Z$. The coordinates of any further point $P$ may be calculated, using the Euclidean metric, from the four distances $OP$, $XP$, $YP$, $ZP$. Obtaining the coordinates for $P$ and $Q$ in this way does not require us to measure the distance $PQ$, yet this distance can be calculated from the coordinates of $P$ and $Q$. Thus, the coordinate description *predicts* this distance. Since the coordinate description can infer the whole *d*-list from a small portion of it, the coordinates must correspond to reality in some deeper way than the *d*-list.

These arguments for 'coordinate realism' surely apply also to probability amplitudes. Consider, for instance, the fact that the state vector generates the chances of possible outcomes for *every* measurement operator. It does not do this in the form of a list of probabilities (a *P*-list, say), but rather by use of a metric on the Hilbert space. The probability of the state $|\psi\rangle$ going to the state $|a\rangle$ after measurement of $A$, for example, is $g^2(|a\rangle, |\psi\rangle)$ or $|\langle a|\psi\rangle|^2$. The transition probabilities, in other words, are not listed in the wavefunction but *generated* using the metric from the probability amplitudes. The complex amplitudes, therefore, seem to represent the system in a deeper way than do the probabilities.

Perhaps the best illustration of how the probability amplitudes seem deeper than the probabilities is the famous two-slit experiment. Let us consider a single particle that passes through a membrane and then has its position measured when it hits a screen. The membrane has a pair of slits cut into it, which are separated by a distance of a few (perhaps 300) wavelengths for the particle. Let us suppose that the particle is initially described by a wavefunction that is a plane wave in physical space.[3] This plane wave passes through the membrane in just the same manner as any classical wave, so that only the parts of the wavefront that pass through the slits manage to get through. These parts undergo diffraction, in the usual way, so that when they reach the screen there is a considerable overlap, and the wave intensity at any point $\boldsymbol{x}$ of the screen may be written as $\psi_1(\boldsymbol{x}) + \psi_2(\boldsymbol{x})$, where $\psi_1$ and $\psi_2$ are the components of the wavefunction due to slits 1 and 2 respectively.

If $\psi_1$ and $\psi_2$ were probabilities, they would have to lie in the interval [0, 1], and so the chance of detecting the particle at $\boldsymbol{x}$ with both slits open would always be greater than for a single-slit membrane. Opening the second slit gives the particle an additional way to get to point $\boldsymbol{x}$. In this case the chance distribution for the particle's position at the screen would be smooth, since the chance distribution for each single slit is smooth. In QM, however, $\psi_1(\boldsymbol{x})$ and $\psi_2(\boldsymbol{x})$ are complex, and it is $|\psi_1(\boldsymbol{x}) + \psi_2(\boldsymbol{x})|^2$ that represents physical chance. It then follows that the chance of the particle being at $\boldsymbol{x}$ when both slits are open may be *less* that the chance for a single slit, as $|\psi_1(\boldsymbol{x}) + \psi_2(\boldsymbol{x})|^2$ may be less than $|\psi_1(\boldsymbol{x})|^2$. (This occurs, for instance, if $\psi_1(\boldsymbol{x}) = \exp(i\pi)$, $\psi_2(\boldsymbol{x}) = \exp(i\pi/6)$, so that $|\psi_1(\boldsymbol{x})|^2 = 1$ but $|\psi_1(\boldsymbol{x}) + \psi_2(\boldsymbol{x})|^2$ is only about 0.54. Such destructive interference occurs when the relative phase of the two components is close to $\pi$.) Also note that, when $\psi_1(\boldsymbol{x})$ and $\psi_2(\boldsymbol{x})$ are in phase, the chance is *four* times what it would be for a single slit.

The chance function for the particle's position at the screen, in other words, shows an interference pattern. The separation of the fringes is determined partly by the wavelength of the particle, which is determined by its energy according to the Planck relation $E = h\nu$. The derivation of this interference pattern within QM depends essentially on the $\psi$-function being a complex probability amplitude rather than a real-number probability. As with the energy levels in the hydrogen atom, it does not make sense to think of 'waves of knowledge' as interfering with each other in this way. Epistemic probability functions, for instance, do not exhibit such interference.

We see that, while it is true that the usual meaning given for the complex amplitudes in the wavefunction is that they generate epistemic probabilities, it seems that they correspond to reality in some deeper way.

### *7.1.2 The Epistemological Interpretation*

The epistemological view of the wavefunction is one of the tenets of the Copenhagen interpretation, at least in my reconstructed version of section 6.1.3. This idea has not yet been developed as clearly as the realistic interpretations of the state vector, such as those of Bohm, Everett, and Penrose, but in this section I will outline and defend the epistemic view as best I can.

The epistemological view of the state vector is essentially that it represents knowledge of a physical system. This knowledge is in the form of providing an epistemic probability for every possible outcome of every kind of measurement that might be performed on the system. One must be careful to distinguish the epistemic view of the wavefunction from what might be called the *statistical*, or *ensemble*, view associated with Einstein.[4] The statistical view is that the state vector represents not a single system but an ideal ensemble of similarly prepared systems. QM is thus similar to statistical mechanics, which describes a system using a coarse-grained partition of the state space into macro-states, defined by large-scale properties such as temperature and pressure. On this view, therefore, it is at least in principle possible to find a more complete representation of a system than its state vector.

We see that there are definite similarities between the epistemic and statistical interpretations, since in both the wavefunction represents a state of knowledge. Indeed, the statistical interpretation really is one (rather mild) kind of epistemic interpretation. As is well known, however, Einstein and Bohr had some fundamental disagreements about QM, so what are the differences? The essential point of contention is that, for Bohr, the information represented by the wavefunction is *maximal*.[5] It says everything about the system that can meaningfully be said. In the statistical interpretation, by contrast, the wavefunction, and the probabilities it generates, are at least largely[6] a reflection of our ignorance of the exact microscopic state of the system. Suppose, for instance, we are about to measure the position of an electron whose wavefunction is not an eigenstate of position. In this case the probabilities of measurement outcomes, which the wavefunction provides, are due to our ignorance of the exact position of the electron. On the epistemic view the probabilities are not due to ignorance, exactly, since on this account there is no further information about the system that one is ignorant of. They are the epistemic probabilities of someone whose information is maximal.

The idea that the wavefunction represents maximal information leads to Bohr's and Heisenberg's view that the limits of measurement accuracy are also limits on *meaning*. The Heisenberg uncertainty principle, for instance, on the orthodox view, does not merely place a limit on the accuracy to which non-commuting observables may simultaneously be measured, but restricts how accurately they may be *defined*. According to the statistical view, by contrast, 'the physical implication of the uncertainty principle is that no state preparation procedure is possible that would yield an ensemble of systems identical in all of their observable properties' (Ballentine 1970: 361). Every individual electron has a well-defined position and momentum at all times; the uncertainty of a quantity describes its statistical distribution across the ensemble.

The key to making at least some sense of this is to remember Bohr's idea that in the quantum domain 'no sharp distinction can be made between the behaviour of the objects themselves and their interaction with the measuring instruments' (Bohr 1949: 42). As Schrödinger (1935: 823) more helpfully put it, in explaining Bohr's 'epistemological viewpoint': 'We are told that no distinction is to be made between the state of a natural object and what I know about it, or perhaps better, what I can know about it if I go to some trouble.'[7] The only way to describe (and define quantities for) microscopic objects is in terms of their interactions with classical, macroscopic objects, and their probabilities of causing certain measurement outcomes. The best description of the quantum world coincides with what we (or someone) can know about that world, from probing it with instruments.

On the epistemic view, therefore, although the probabilities of quantum mechanics are epistemic probabilities, representing information, they are also ultimate, or irreducible. They are irreducible since they represent the *best possible* knowledge of the system; there is no more complete description to be had. Being, it appears, something of a positivist, Bohr finds no meaning to the word 'reality' beyond what can in principle be described, and so identifies reality with what Schrödinger (1935: 823) calls *Meßwirklichkeit*, or measurement reality. Wigner (1963: 337) describes Bohr's view that the probabilities of QM are ultimate as follows:

> We recognize, from the preceding discussion, that the state vector is only a shorthand expression of that part of our information concerning the past of the system which is relevant for predicting (as far as possible) the future behavior thereof. The density matrix, incidentally, plays a similar role except that it does not predict the future behavior as completely as does the state vector. We also recognize that *the laws of quantum mechanics only furnish probability connections between results of subsequent observations carried out on a system*. It is true, of course, that the laws of classical mechanics can also be formulated in terms of such probability connections. However, they can also be formulated in terms of objective reality. The important point is that the laws of quantum mechanics can be expressed only in terms of probability connections.

It is important to bear in mind that the only quantum reality that Bohr recognizes, namely, the *Meßwirklichkeit*, is something quite objective and independent of the mind. Bohr constantly stresses, for instance, that an observation is finished as soon as the system has interacted with a measurement apparatus. The recorded value exists in the device whether or not a conscious observer ever looks at it, so that he does not share the position of Wigner (1961) or Wheeler (1981). This objective status for the wavefunction clearly requires that it represent maximal, or best possible, information about the system, the limits of which are set by the actual physical nature of interactions. Otherwise it would depend upon, for example, the diligence or laziness of the experimenter in the trouble to which he goes in learning about the system.

I realize that this account of the epistemic view is rather vague, but it is an accurate summary of the clearest discussions available. Fortunately, as will be shown below, these ideas can be made more precise using the causal theory of chance. For now, let us look at the arguments that motivate such an epistemic view of the $\psi$-function.

The first point to note is that the $\psi$-function is defined not on physical space but rather on the configuration space for the system. The number of dimensions of this space is the number of degrees of freedom of the system concerned, and so may be much greater than the three dimensions of physical space. For a system of $N$ particles, for instance, the configuration space has $3N$ dimensions, and there is a single, shared wavefunction for all the particles together. This does not itself mean that the wavefunction is not physical, of course. In Hamiltonian mechanics, for example, the state of a system is represented by a point in the phase space, which has $6N$ dimensions for an $N$-particle system, but there is no temptation to say that classical states are epistemic. For one thing, each point in phase space is equivalent to a specification of the position and momentum for each of the $N$ separate particles, so that the phase space can (in principle) be dispensed with, in describing the state of the system.

The significance of the use of configuration space in quantum mechanics arises from the fact that physical quantities are, in general, 'smeared out.' That is to say, a wavefunction $\psi$ for a physical system does not (in general) specify a definite value for each quantity $A$. Rather, it gives a set of possible values for $A$, such as $a_1$, $a_2$, and so on, each of which has a complex amplitude $c_1$, $c_2$, etc. This smearing out of physical quantities prevents the wavefunction for a composite system from being split up into separate wavefunctions for the component subsystems. As an example of this, consider the positions of a pair of particles **X** and **Y** in one dimension. If **X** has position coordinate $x$, and **Y** has position $y$, then the total system $\langle \mathbf{X}, \mathbf{Y} \rangle$ has a single wavefunction $\Psi(x, y)$. In general, this cannot be expressed as a product of wavefunctions, one for each system, that is, $\Psi(x, y) = \psi_X(x)\psi_Y(y)$. This is in sharp contrast with classical mechanics, where the state of a composite system can always be decomposed into a set of states – one for each subsystem.

In the general case where $\Psi$ does not factorize into separate wavefunctions for **X** and **Y**, the complex number $\Psi(x, y)$ is not associated with any point in space – it is not *localized*, in other words. The quantity $\Psi(x, y)$ is instead associated with *two* points in the same space, $x$ and $y$. This is unusual, as we expect physical properties to be located in space and time, not floating free. What would one make of a temperature field that assigned a temperature to each *pair* of points on a metal rod, in such a way that it could not be reduced to a pair of localized fields? What interpretation could be given to the number $T(x, y)$? It would be a temperature that did not exist anywhere in physical space.

Non-localized quantities are therefore difficult to interpret as representing physical properties. On the other hand, we have already seen in chapter 6 that *information* is very frequently non-localized. A simple example of this was the case of the two coffee mugs, where it was known that one of them contained sugar. This information about the pair of mugs was not localized, in the sense that it could not be split up into separate pieces of information about each mug. The epistemic probability function generated by this information is also not localized, in the sense that each probability attaches irreducibly to a joint event, whose parts are spatially separated. Thus, the mere fact that the wavefunction is not localized strongly suggests an epistemic reading.

One may counter this reasoning by saying that we should simply reject the notion that physical properties must be localized in spacetime. Why cannot we regard the configuration space as physically real, in some sense? Indeed, this is precisely the view held by David Bohm, for example. In response to this suggestion, we can make the following points:

1 We should not posit such a space unless there are strong advantages to doing so, according to Ockham's Razor. The epistemic interpretation is far more parsimonious.
2 If we do allow non-localized properties to be physical, then we must also admit *causal* non-locality, or action-at-a-distance.

The second point is clear from the fact that the wavefunction, though not localized in space, does provide probabilities of measurements made at specific locations. Moreover, if the wavefunction is physical, then the probabilities it generates must also be physical. Now, we saw in the EPR experiment that the probability of an experimental outcome in one place can be altered by carrying out an experiment somewhere else, on a different system. If these probabilities are physical, then this constitutes action-at-a-distance, as one is bringing about a physical change at a distant location.

It would be rash to rule out causal non-locality as impossible, but any theory that requires causal non-locality is hard to accept. Any alternative account that preserved causal locality would (*ceteris paribus*) be greatly preferable. But such an alternative is readily available, as we have seen in section 6.3. The failure of joint probability distributions for composite systems to factorize is easily understood within the causal theory of chance, where chance is an epistemic probability. It arises directly from the fact that maximal information about the composite system may not be localized.

It should be noted that this epistemic understanding of QM correlations is *not* available to the statistical interpretation. This naive epistemic view involves the claim that each observable on a single system possesses a definite value, which (locally) determines the outcome of a measurement of it. As we saw in the previous chapter, however, this is precisely the assumption used to derive the Bell inequality. Thus, within the statistical interpretation, the mere correlation of knowledge will not produce the required degree of correlation in the observed frequencies. One may try to fill the gap by postulating some non-local causal interactions, but then the advantage of an epistemological understanding of QM correlations has been given up.

### *7.1.3 The AND/OR Problem*

The AND/OR problem concerns the two different ways of reading superpositions of pure states, which are most vividly exposed in connection with delayed-choice experiments. Consider, for instance, the experiment described by Wheeler (1981: 183), as displayed in figure 7.1.

A beam of light hits the half-silvered mirror $M_1$ on the bottom left, so that half of it is reflected and half passes through. The two halves of the split beam take different routes through the apparatus, going via the mirrors A and B, but they converge again at a second half-silvered mirror $M_2$ just in front of the counters $D_1$ and $D_2$. This mirror reflects only half of the light from A, transmitting the rest, and the same for the light from B, so that each of $D_1$ and $D_2$ receives a superposition of two beams. The geometry is set up so that the two beams entering $D_1$ are in destructive interference and cancel out, whereas the beams entering $D_2$ interfere constructively. Thus, only $D_2$ registers the arrival of any photons.

This result, that only $D_2$ records the impact of photons, holds even when the intensity of the light is so low that there is only one photon in the apparatus at a time, so it may be difficult to understand how there can be any interference.

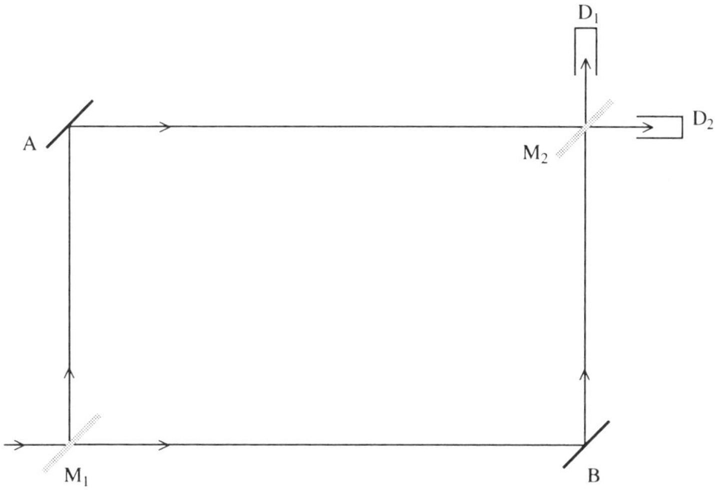


Figure 7.1

Each photon, one would think, is sent by $M_1$ either up to mirror A or across to B. Then, among those photons arriving from A, half will end up at $D_1$ and half at $D_2$, and the same for photons arriving from B. Thus, the overall counts by $D_1$ and $D_2$ should be roughly equal.

Let us look at how QM treats this experiment. When a photon hits $M_1$, its wavefunction becomes a sum of two components, one of which heads for mirror A and the other for B. It is as if even the single photon is split into two, with one part taking each route. These two components, when they strike the mirror $M_2$, each split again so that each of $D_1$ and $D_2$ gets a sum of two 'quarter' wavefunctions. For $D_1$ these two components are out of phase and so cancel out, whereas for $D_2$ they reinforce each other. QM thus obtains the correct result, that photons are registered only at $D_2$, by allowing the photon's wavefunction to travel both routes, and so to interfere with itself.

The 'delayed choice' aspect of this experiment concerns the fact that, if we leave out the second half-silvered mirror $M_2$, counters $D_1$ and $D_2$ receive roughly equal numbers of photons. In this case one *might* think that each photon takes a definite route through the apparatus, but this cannot be so. For the choice of whether or not to put $M_2$ in place may be delayed until the very last moment, so that a photon 'looking ahead' and seeing that $M_2$ is missing, and thus deciding to take a definite route, would be foiled by a last-moment insertion of $M_2$. Such a route-definite photon might easily be recorded at $D_1$, but this never happens in actual experiments. Even when $M_2$ is missing there can still be no definite route for the photon, since it has to be prepared for the possibility that $M_2$ be added later.

Whether or not the mirror $M_2$ is present, therefore, the photon's wavefunction is a sum of two components, which we shall write schematically as $|a\rangle$ + $|b\rangle$, for the parts that travel via mirrors A and B. It seems that, in view of the possible interference between them, we must regard each of $|a\rangle$ and $|b\rangle$ as representing something physically real, rather than just an abstract possibility. As Bell (1990: 36) helpfully puts it, the elements of the superposition *coexist* in the concrete world, so that we have $|a\rangle$ AND $|b\rangle$.

Now let us consider what happens to the wavefunction after it reaches the detectors, in the case where the mirror $M_2$ is not put in place. According to QM the photon becomes correlated with the detectors, so that the final state may be written as

$$|\Psi\rangle = |a\rangle|-\rangle_1|+\rangle_2 + |b\rangle|+\rangle_1|-\rangle_2,$$

where $|-\rangle_1$ and $|+\rangle_1$ are the states of $D_1$ corresponding to not detecting and detecting the photon, respectively, and $|-\rangle 2$ and $|+\rangle 2$ similarly for $D_2$. One generally assumes[8] that each measurement has a definite outcome, so that each of $D_1$ and $D_2$ either has definitely detected a photon or has definitely not detected a photon (a detection being indicated perhaps by a numeral displayed on an LCD increasing by one). If each counter has a definite state, however, then it seems that only one of the elements of the final superposition is physically real, the other being a mere abstract possibility. In other words, we read the superposition as $|a\rangle|-\rangle_1|+\rangle_2$ OR $|b\rangle|+\rangle_1|-\rangle_2$.

This is what I call the AND/OR problem in interpreting the state vector. In some circumstances we read superpositions as AND, but in other cases we read them as OR. It is problematic since the connectives AND and OR are fundamentally distinct, and so this variation of meaning looks like a nasty equivocation. This problem is sometimes called the problem of the 'collapse' of the wavefunction. The reading of a superposition as OR is equivalent to applying von Neumann's Projection Postulate, causing a superposition to collapse into just one of its elements. I consider the AND/OR talk to be more helpful, however, as the term 'wavefunction collapse' suggests a physical change whose status is questionable at best.

The usual rule for deciding whether to read a superposition as AND or OR is roughly as follows: a superposition is read as OR whenever a measurement is deemed to have occurred, and as AND at all other times. This is problematic, however, for two reasons. First, although the term 'measurement' is clear enough for most practical purposes, it has no theoretically precise definition.

Second, even if a precise definition could be found, it should not merely demarcate measurements from non-measurements but also explain *why* measurements deserve such special treatment. These problems are discussed in more detail in section 7.1.5 below.

How are the AND and OR readings of superpositions related to the physical and epistemic interpretations of the state vector? Let us first consider the physical interpretation, that the wavefunction represents a physical field of some kind. If a field has different values at different points in space, then we do not regard these as alternatives, but as all existing together in the real world. The field has value $c_1$ at $x_1$ AND $c_2$ at $x_2$ AND $c_3$ at $x_3$. The weights $c_1$, $c_2$, $c_3$ represent physically real properties, not degrees of belief. On this view, therefore, if a particle's wavefunction is a superposition of position eigenstates $|x_1\rangle$, $|x_2\rangle$, $|x_3\rangle$, we would say that the particle (or something physically real associated with the particle) exists at all three places together.

The physical interpretation easily accounts for the AND reading of superpositions therefore, but the OR reading is less straightforward. In Bohm's account the particle trajectories come to the rescue, by effectively selecting one position eigenstate. For Everett the OR interpretation is an illusion, since in fact all the elements of the superposition exist. According to others, such as R. Penrose, and Ghirardi, Rimini, and Weber, the vanishing of all but one element of the superposition is explained in terms of a physical change.

Now let us consider the epistemic interpretation, in which the wavefunction is closely connected to epistemic probability. When each of a number of distinct alternatives has a certain epistemic probability, only *one* of these possibilities actually obtains. If, for instance, the unseen card in my hand has epistemic probability 0.2 of being an ace, 0.4 of being a king, 0.3 of being a ten, and 0.1 of being a jack, then it is either an ace OR a king OR a ten OR a jack. These alternatives all exist together in the epistemic probability function, side-by-side so to speak, each with a certain weighting, but this a just a matter of knowledge. In fact – that is, in the real world – only one of these alternatives exists. Thus, when the wavefunction assigns non-zero amplitudes to a number of distinct alternatives, such as to position eigenstates $|x_1\rangle$, $|x_2\rangle$, and $|x_3\rangle$, from the epistemic point of view it would seem most natural to say that the particle either has position $x_1$ OR $x_2$ OR $x_3$.

This reasoning ignores the fact, discussed below in §7.2.4, that the term 'measurement' does not have its usual meaning in quantum mechanics. The result of a quantum measurement does not, in general, indicate the value of a pre-existing quantity in the system. Thus, if a system has state $|\psi\rangle = c_1|x_1\rangle + c_2|x_2\rangle + c_3|x_3\rangle$, we cannot infer that the system is really in just one of the positions $x_1$, $x_2$, or $x_3$. If measured, it would be *found* in one of those positions, but that position did not exist prior to the measurement. Moreover, this view of superpositions as disjunctions is ruled out by Bohr's claim that the state vector represents maximal information about the system. If, for instance, only the term $|x_3\rangle$ existed in reality, then the state $|\psi\rangle = |x_3\rangle$ would contain more information than $|\psi\rangle = c_1|x_1\rangle + c_2|x_2\rangle + c_3|x_3\rangle$, and so the latter would not be maximal.

It follows that, when a measurement is not being made, all three terms of the superposition have to be taken seriously, each as a necessary part of the best possible representation of the system. This is not exactly an AND reading, as it does not claim that the terms each correspond to something physically real, but it is close to an AND reading.

It should be noted that, according to the epistemic interpretation, the Schrödinger equation does not apply to all processes without exception. By reading a superposition as OR, after a measurement, one assumes that the wavefunction does not represent maximal information at that time. The OR reading supposes that the physical world contains a result for the measurement, in the form of a definite final state for the apparatus, yet this result is not contained in the final wavefunction. The wavefunction in this case, failing to describe something that is there to be described, is clearly sub-maximal.

We thus reach the conclusion that, for the OR reading of a superposition (after measurement) to be valid, within the epistemic interpretation of the state vector, the state vector does not always represent maximal knowledge. This loss of maximality would be due, presumably, to the measurement process being stochastic rather than deterministic. It is part of the very definition of a stochastic process that maximal knowledge of that process at one time does not allow one to infer exactly what will happen at a later time, so that maximal knowledge about $t_1$ does not provide maximal information about a later time $t_2$.

### *7.1.4 The Shifty Split*

I have avoided the terms 'collapse of the wavefunction,' 'reduction of the state vector,' and so on since I consider them unhelpful. They suggest some sort of physical change that corresponds to the replacement of a state vector with the eigenstate of some observable (or with a statistical mixture of such eigenstates), yet it is doubtful that any such physical process exists. Indeed, even among those who propose a purely physical interpretation of the state vector, only R. Penrose and GRW recognize something like the wavefunction collapse as a physical process. According to both Bohm and Everett, the Schrödinger equation applies universally.

In my terminology, the reduction of the state vector is (roughly speaking) just the shift from the AND to the OR reading of the state vector. Instead of seeing the elements of a superposition as existing together in the real world, we consider that only one of them is really there, and interpret the squared modulus of each coefficient as its epistemic probability of being the real one. We must be more precise than this, however, since any pure state can be expanded as a superposition using many different alternative orthonormal bases. A ket vector $|\psi\rangle$, for instance, may be expanded in either of the bases $\{|a_i\rangle\}$ or $\{|b_i\rangle\}$, but it is not possible to read both of these superpositions as OR. In that case one would think the system really had some state $|a_k\rangle$, and also some state $|b_r\rangle$, but these are incompatible in general. The OR reading can only be applied to one superposition at most, that is, with respect to only one complete orthonormal basis. In practice the base vectors used are generated by eigenstates of 'macroscopically-local' observables.[9]

Since the AND reading of the wavefunction is applied to some systems, while the OR reading is used for others, quantum mechanics involves a split between the parts of the world to which each reading applies. On one side of the split is the 'system,' whose superpositions are all read as AND, and on the other side is the 'apparatus,' the 'environment,' and so on, *some* of whose superpositions (those involving suitable observables) are read as OR. The OR reading applies not only to the wavefunction of the apparatus itself, but also to any joint wavefunction for the system and apparatus together. As Everett's theory highlights, an orthonormal basis for the apparatus defines a unique decomposition of the state vector for the total system+apparatus, where each term of the superposition is the product of a base vector for the apparatus and the corresponding relative state for the system.

We saw in the discussion of the delayed-choice experiment that the OR reading of the state vector can be applied inappropriately. After the photon has encountered the first mirror $M_1$, for example, its state vector can be expressed as a sum of two components, which take different routes through the apparatus. If we apply the OR reading at this point, and regard one of the components as fictional, then we cannot account for the experimental results actually obtained, which require interference. Is it also possible to apply the AND reading inappropriately, and so predict the wrong empirical results? In principle this is possible, although it is virtually impossible in practice, as the following results show. The move from AND to OR can be delayed as long as one desires, without affecting (FAPP)[10] the empirical predictions finally obtained. Of course, to get empirical predictions (probabilities of measurement outcomes) from the theory, we must eventually apply the OR reading. The first of these results is due to von Neumann (1932: 641–4).

Let us divide the world up into three parts: a system **X**, a measurement apparatus **M**, and the rest of the world **R**. The system **X** is described quantum-mechanically, and so has some state vector $|\psi\rangle_{\mathbf{X}}$. Suppose we make a measurement on **X**, using **M**. If the AND/OR split is applied at the interface between **X** and **M**, so that ⟨**R**, **M**⟩ is described classically, then this measurement results in $|\psi\rangle_{\mathbf{X}}$ jumping to one of the eigenstates of the operator $A$ measured by **M**. There is no obvious need to place the split at the **M** − **X** interface, however, since **M** is made of atoms, and so it should be possible to describe it also using quantum mechanics. Perhaps we can instead divide the world into **R**, treated classically, and ⟨**M**, **X**⟩, which is given a state vector? In this case, when **X** interacts with **M** during the measurement, the Schrödinger equation applies throughout and the two systems become correlated. A definite measurement result then does not emerge until something in the environment **R** measures the measurement outcome of **M**.

The important question is whether or not the formalism of QM allows the AND/OR split to be shifted in this way. In other words, are these two procedures consistent, in the sense of providing the same probabilities for measurement results? Von Neumann showed that they are consistent, as follows.

Suppose first we divide the world into ⟨**R**, **M**⟩ and **X**. Then the possible results of the measurement of **X** by **M** are the eigenvalues $a_i$ of the operator $A$, with probabilities $|\langle a_i|\psi\rangle|^2$. If, however, we treat **M** quantum mechanically, then the state of ⟨**X**, **M**⟩ before the measurement is

$$\begin{aligned} |\Psi\rangle &= |\psi\rangle_{\mathbf{X}}|\varphi_0\rangle_{\mathbf{M}} \\ &= \left(\sum_i c_i|a_i\rangle_{\mathbf{X}}\right)|\varphi_0\rangle_{\mathbf{M}}, \qquad (1) \end{aligned}$$

where $|\phi_0\rangle_{\mathbf{M}}$ is the initial wavefunction for **M**. Now suppose that **X** is initially in some eigenstate $|a_k\rangle$ of $A$, that is, $\psi = |a_k\rangle$. In this case we want ⟨**X**, **M**⟩ to evolve, during the measurement, to a state such as $|a'_k\rangle_{\mathbf{X}}|b_k\rangle_{\mathbf{M}}$, where $|b_k\rangle_{\mathbf{M}}$ is an eigenstate of some macroscopically local property $B$, such as the position of a needle, and $|a'_k\rangle_{\mathbf{X}}$ is the final state of **X**.[11] (We assume that $\langle b_j|b_k\rangle_{\mathbf{M}} = \delta_{jk}$.) This is the criterion of a successful measurement interaction. If such a measurement occurs when **X** is in a superposition of eigenstates of $A$, as in (1), then since the time-evolution operator is linear the final state will be

$$|\Psi\rangle = \sum_k c_i|a_i'\rangle_{\mathbf{X}}|b_i\rangle_{\mathbf{M}}.$$

If we now apply the OR reading to ⟨**X**, **M**⟩, then the probability of some measured value $b_k$ on the measurement apparatus **M** is

$$
\begin{aligned}
P(B = b_k) &= \sum_i P(A = a_i \ \&\ B = b_k) \\
&= \sum_i \left| \langle b_k|_{\mathbf{M}} \langle a_i|_{\mathbf{X}} \left( \sum_j c_j |a_j'\rangle_{\mathbf{X}} |b_j\rangle_{\mathbf{M}} \right) \right|^2 \\
&= \sum_{i,j} \left| c_j \langle b_k \,|\, b_j\rangle_{\mathbf{M}} \langle a_i \,|\, a_j'\rangle_{\mathbf{X}} \right|^2 \\
&= \sum_{i,j} \left| c_j \delta_{kj} \langle a_i \,|\, a_j'\rangle_{\mathbf{X}} \right|^2 \\
&= |c_k|^2 \sum_i \left| \langle a_i \,|\, a_k'\rangle_{\mathbf{X}} \right|^2 \\
&= |c_k|^2
\end{aligned}
$$

Therefore, the probability of each result $A = a_k$ and $B = b_k$ is $|c_k|^2$, that is, $|\langle a_k \,|\, \psi\rangle|^2$, and so the two approaches are consistent.

This result is not surprising in view of the property of correlated systems shown in §6.1.2: that when two systems **X** and **Y** are correlated each has a definite state relative to the other. More precisely, for each state $|\psi\rangle_{\mathbf{X}}$ in the subspace for **X** there is a unique corresponding relative state for **Y**. Also, both **X** and **Y** have density matrices, which means that a measurement on **X** alone (or on **Y** alone) has exactly the statistics that it would have if **X** (or **Y**) had a definite, but unknown, pure state. In short we may say that, if we can subsequently only measure one of the systems concerned, a correlation between **X** and **Y** is indistinguishable from a measurement of **Y** by **X**. The 'collapse of the wavefunction' seems to have something to do with correlation.

Von Neumann's theorem only shows that measurements on **M** by **R** cannot distinguish a **M** − **X** correlation from **M** having a definite state. If **R** measures both **M** and **X**, on the other hand, and compares the results, then (in principle at least) one can determine whether or not the wavefunction was collapsed by the measurement of **X** by **M**. Von Neumann considers an ascending 'chain' of measuring devices, stretching from the system to the conscious observer, where each device measures only the outcome of the previous instrument. In such a situation it is impossible to tell at what point in the chain the OR reading becomes valid, or a definite outcome emerges. If a device is able to measure *all* the systems that may be correlated together, however, then it is possible to see if the correlation is really there.

The practical problem with such an experiment arises from the fact that measuring instruments are large objects, which under normal circumstances are impossible FAPP to isolate from the environment. For suppose that, upon measurement of **X** by **M**, the two merely become correlated, that is, the wavefunction does not collapse. Since **M** is macroscopic, however, it is in constant interaction with the environment **R**, and so ⟨**X**, **M**⟩ very quickly becomes correlated with **R**. Thus, even measurements on both **X** and **M** will give the appearance of a wavefunction collapse. To see any interference effects we must measure ⟨**X**, **M**, **R**⟩, which is impossible FAPP. That would involve measuring billions of systems together, at the least.

The elimination of coherence FAPP by interactions with the environment seems to provide a solution to the *practical* problem of when to shift to the OR interpretation of the state vector. Whatever the 'wavefunction collapse' may be, and at whatever stage it may occur, is of no practical concern since the onset of decoherence brings about a reduction FAPP. Indeed, as Joos and Zeh (1985) show, many at least of the familiar classical properties of systems, such as the well-defined spatial structures of all but the smallest molecules, depend essentially on the destruction of phase coherence *in the system* by interaction with the environment.[12] It is important to realize however, that such work, while of great importance, cannot succeed in giving a complete derivation of classical concepts (the OR reading) from quantum mechanics (the AND reading). As Joos and Zeh explain, 'The use of the local density matrix allows at most only a partial derivation of classical concepts for two reasons: it already assumes a local description, and it presupposes the probabilistic [i.e., OR] interpretation leading to the collapse of the state vector at some stage of a measurement' (1985: 224).

Some have suggested that decoherence provides not just a solution to the practical problem of when to shift from AND to OR, but rather a complete reconciliation of this difference. As Stamp puts it, 'It is then argued that there is no longer any measurement problem or paradox, and that the classical world of Bohr is *defined* as the world of phenomena in which decoherence has already taken place (1995: 127).' This idea dates back at least to Jordan (1949) and, as Stamp notes, 'it seems to get reinvented every five years or so' (1995: 127). Concerning this proposed solution of the AND/OR problem I share the view of John Bell, that it is very hard to see how it is supposed to work. Like Bell, I do not see how '... the elimination of coherence, in one way or another, implies the replacement of "and" by "or"' (1990: 36).

### *7.1.5 The Measurement Problem*

In orthodox interpretations of quantum mechanics, measurement is forced to play a fundamental role. As shown in the previous section, for example, the concept of measurement is used to create a shifty split between 'system' and 'apparatus,' which are the spheres of application for the AND and OR readings of the state vector. According to Bohr, the role of the measurement apparatus in QM is similar to that of the reference frame in relativity theory, as a necessary vantage point from which to describe the world (Bohr 1935: 150–1). The quantum-mechanical description is therefore the world as seen from the point of view of a measurement apparatus. (In postmodern terminology, one might say that QM is *metrocentric*!)[13]

Many of us find the metrocentric nature of quantum mechanics problematic. Measuring instruments are, after all, just ordinary physical systems, made of atoms, and measurement interactions are ordinary physical interactions. Thus, the fundamental theoretical role of measurement within orthodox QM is a problem, known as the measurement problem or paradox. The measurement problem is closely connected to the tension between the physical and epistemological interpretations of the state vector. On the physical interpretation of the wavefunction it is only to be expected that measuring instruments will have a special epistemological status, since they provide us humans with knowledge. They should not, however, have any special *physical* status. On the epistemological interpretation of the ψ-function it is less clear what role measuring instruments should have, but it is likely to be quite fundamental.

Perhaps the most famous critic of standard QM, and its special role for measurement, is John Bell. Bell complains that 'surely, after 62 years, we should have an exact formulation of some serious part of quantum mechanics?' (1990: 33). By 'serious,' he means partly that the '"apparatus" should not be separated off from the rest of the world into black boxes, as if it were not made of atoms and not ruled by quantum mechanics' (33). Bell is especially dissatisfied with the fact that standard QM is only concerned with the results of measurements, that it seems to see the world only from the point of view of a measuring instrument. A satisfactory theory would be concerned with reality, not just *Meßwirklichkeit*; it would describe not *observ*ables but *be*ables (1973: 41, 1975). In short, we may say that Bell wants to describe reality from God's point of view, not from a measuring instrument's point of view. Such a divine description would, for example, be able to talk about the measurement process itself from beginning to end.

The kind of solution to the measurement problem that Bell wants, therefore, is a new formulation of QM, or a new theory altogether,[14] in which measuring instruments have no special theoretical status. In such a theory the measurement process would be treated in the same way as any other interaction, so that measurement is 'dethroned.' Bell is encouraging toward attempts to provide such a theory, particularly those of Bohm and GRW. (He regards Everett's ideas less favourably, however.)[15]

A second kind of solution to the measurement problem would be to keep measurement in its dominant place, but to show that this role is legitimate and necessary. Such a solution must provide a theoretically rigorous definition of measurement that satisfies two criteria. First, it must draw a line between measurements and other interactions in a way that both makes physical sense and is consistent with practice. Second, it must explain *why* the split between system and apparatus is legitimate and necessary.

At the present time there is no account of measurement that meets these criteria. Bohr, for instance, regarded the properties of amplification and irreversibility[16] as important features of a measurement, but did not give a precise analysis. A measurement apparatus is a macroscopic system, being composed of $10^{24}$ or so atoms, and thus a measurement is the right kind of interaction with such a 'large' system. Also a measurement interaction must be irreversible, as it results in a permanent record such as a photograph. Thus, since thermodynamic irreversibility exists only for systems with many degrees of freedom, a measurement must involve an 'amplification'[17] to the macroscopic level.

These ideas of Bohr are of some use in demarcating measurements from other interactions. In the delayed-choice experiment, for example, the interaction between the photon and the mirror $M_1$ does not involve any amplification to the macroscopic level, so it should not count as a measurement. When the photon reaches the detectors, by contrast, it is a different story. The problem is that these ideas do not even begin to satisfy the second criterion, of showing why measurement has to play such a central role.

## 7.2 LARGE AND SMALL SYSTEMS

In the discussion of factorization in section 6.3.1 of the previous chapter, I argued that a description of a pair of systems is more likely to factorize if it concerns large-scale properties of large systems. Moreover, in this chapter we have seen that the epistemological interpretation of the state vector also appeals to a distinction between large and small systems. The probabilities generated by the state vector are probabilities of measurement outcomes, which must be events 'at the macroscopic level.' In sections 7.2.1 to 7.2.3 I shall argue that this appeal to size is valid.

The appeal to size in the very *formulation* of quantum mechanics certainly looks highly dubious. It is not a problem that large and small systems behave differently, of course, as this is even true in classical mechanics. For example, in the atmosphere a large particle in free fall has a greater terminal speed than a small one made of the same material, but this difference is explicable within a general theory that applies to particles of all dimensions. (As one increases the radius of a sphere, its volume increases more rapidly than its surface area, so the weight increases faster than the drag.) The problem with QM is, rather, that large and small systems seem to follow fundamentally different rules, which cannot be unified by deriving them both from a more general account. An electron, for instance, does not always have a definite velocity, but may exist in a superposition of velocity eigenstates. A cricket ball, by contrast, always has a (more or less) definite velocity.[18] The important point is that there is no overall principle that explains why cricket balls, but not electrons, always have definite velocities. If you apply QM strictly to large systems, without arbitrarily collapsing the wavefunction whenever a superposition of macroscopically distinct states develops, then one infers that a cricket ball *can* be in a superposition of macroscopically distinct velocities.[19]

Bell is thus apparently right to include the terms 'microscopic' and 'macroscopic' on his black list of words which 'however legitimate and necessary in application, have no place in a *formulation* with any pretension to physical precision' (Bell 1990: 34).[20] I shall argue, however, that (in some sense) new properties emerge in large systems that are not present in their parts.

### *7.2.1 Emergent Properties*

For brevity, I will sometimes use expressions like 'description of **X**' and 'model of **X**' in place of 'maximal actual state of affairs concerning **X**.' These terms also remind us of the fact that states of affairs are components of epistemic states.

What is meant by 'large' and 'small' systems? We are obviously concerned with a continuous scale here, rather than a binary division, so we should begin by defining a size ordering, that is, by giving the meaning of '**X** is larger than **Y**.' I suggest the following definition.

**7.2.1.1 Definition** System **X** is larger than system **Y** if **Y** is similar to a proper subsystem of **X**.

This definition only gives a partial ordering of course, and so needs to be supplemented,[21] but it expresses the general idea that 'size' in this sense is a matter of *composition* rather than length. A pair of electrons separated by a distance of five miles is large in some sense, but not in the sense used here. It is considered much smaller than, for instance, a one-micron speck of dust, as the speck will contain at least ten billion electrons.

The world revealed by quantum mechanics is strange and unfamiliar, but this is not itself problematic. Indeed, it would surely be odd if protons behaved just like cricket balls, given that the mass ratio between the two is about $10^{27}$! It is very easy to accept that the laws that operate at the atomic level are different from, and not derivable from, the familiar rules of the common-sense world. One feels, however, that the reverse should not be true. In principle at least, it should be possible to derive the rules for large objects as limiting cases of the rules of QM, in something like the way that Newtonian mechanics can be derived from special relativity in the limit as *c* tends to infinity. One should, in principle at least, be able to treat a cricket ball quantum-mechanically, merely as a large collection of atoms that jointly obey the Schrödinger equation.

Why should one be able successfully to apply the rules of QM to large objects? The idea seems to be that a whole system is nothing over and above the sum of its parts, so that there are no 'emergent properties.' Thomas Nagel (1979: 182) expresses this view as follows:

> There are no truly emergent properties of complex systems. All properties of a complex system that are not relations between it and something else derive from the properties of its constituents and their effects on each other when so combined. Emergence is an epistemological condition: it means that an observed feature of the system cannot be derived from the properties currently attributed to its constituents. But this is a reason to conclude that either the system has further constituents of which we are not yet aware, or the constituents of which we are aware have further properties that we have not yet discovered.

It is interesting to note that, in later arguing for this thesis that there are no emergent properties, Nagel treats the parts of an object as *causes* of it. The total system exists because its subsystems do, so that the existence of the subsystems is constitutive of, or supports, the existence of the whole. Indeed, it is on the basis of this causal relation between the parts and the whole that Nagel infers that the properties of the parts must necessitate the properties of the whole, since he holds that causes always necessitate their effects.[22]

In view of the distinction between efficient causation and determination, as relations between events, which I argued for in chapter 3, we may wonder if there is a similar distinction to be made between material causation and determination, as relations between parts and wholes. Indeed, on the face of it there is just such a distinction. On the one hand, there is an ontological relation, whereby the existence of the whole is constituted by, or caused by, the existence of the parts. Also, however, there is the logical relation between the best descriptions of the parts and the best description of the whole. The causal relation holds between concrete systems, whereas the logical relation is defined on possible states of affairs concerning those systems.

We see therefore that we must consider two distinct kinds of emergence, namely, causal emergence and logical emergence. The concrete system **X** is causally emergent if it is not caused, or constituted, by **X**'s subsystems $\mathbf{X}_1$, $\mathbf{X}_2$, ..., $\mathbf{X}_n$. A state of affairs of a whole system **X** is logically emergent just in case it is not logically entailed by the maximal states of affairs concerning **X**'s subsystems $\mathbf{X}_1$, $\mathbf{X}_2$, ...,$\mathbf{X}_n$.[23]

Causal emergence seems to be impossible. The very idea of parts and wholes surely entails that the whole is, ontologically speaking, nothing over and above its parts. I shall argue, however, that logical emergence, the kind that Nagel discusses, is quite possible. This situation is exactly analogous to the case of relations between events: every event has total causes, but it may not be necessitated by its causes. A random event is one that, although fully caused, is not fully determined by those causes. In a similar way, the large-scale nature of a system is always fully caused by the concrete natures of its parts, but its high-level properties may not be logically determined by its low-level properties. We might even say, by analogy, that the high-level properties are *random*.

I have already argued in the previous chapter for the existence of logically emergent states of affairs, although under a different guise. Consider a composite system $\langle \mathbf{X}, \mathbf{Y} \rangle$ whose maximal state of affairs $m\langle \mathbf{X}, \mathbf{Y} \rangle$ does not factorize into separate propositions $m(\mathbf{X})$ and $m(\mathbf{Y})$, concerning **X** and **Y** respectively. Of course the proposition $m\langle \mathbf{X}, \mathbf{Y} \rangle$ must entail $m(\mathbf{X})$ & $m(\mathbf{Y})$, since otherwise it would not be maximal concerning $\langle \mathbf{X}, \mathbf{Y} \rangle$. The entailment fails in the other direction, as $m(\mathbf{X})$ & $m(\mathbf{Y})$ does not entail $m\langle \mathbf{X}, \mathbf{Y} \rangle$. We will write this situation as

$$m\langle \mathbf{X}, \mathbf{Y} \rangle \Rightarrow m(\mathbf{X}) \ \& \ m(\mathbf{Y}).$$ [24]

In this case the high-level proposition $m\langle \mathbf{X}, \mathbf{Y} \rangle$ is, according to the above definition, logically emergent. It cannot be inferred, with certainty, from the best descriptions of the subsystems. Thus, in short, the theory of non-classical correlation from chapter 6 commits us to accepting logical emergence.

It should be obvious that the failure of logical emergence requires that the maximal states of affairs concerning the subsystems be imperfect, or incomplete. If the concrete subsystem were nothing more than its maximal state of affairs, then the conjunction of these low-level propositions would equal the total system.

What kinds of properties are left out by the low-level maximal description? In the discussion of the EPR experiment, we saw that $m\langle \mathbf{X}, \mathbf{Y} \rangle$ contains *non-*

*local* information about the pair of systems, that is, information that does not concern either system individually but only the whole system together. In other words, the extra information at the higher level concerns the large-scale structure of the whole system. In the case of a macroscopic system, some of its large-scale properties will be logically emergent, that is, not inferable even from maximal propositions concerning its atomic-level subsystems. Of course not *all* of its large-scale properties are emergent. Its mass, for instance, which is a large-scale property concerning the whole system together, seems to be just the sum of the masses of its component parts. Thus, its mass may be inferred from a maximal description of its parts. My point is merely that *some* of the large-scale properties of a macroscopic body, those properties that concern the whole body together, cannot be inferred from the properties of the parts.

### *7.2.2 Saturated Models for Large Systems*

As one describes the world at larger and larger scales, does the emergence of new properties ever end? Is there a level such that the description of the world at that level is as complete as those above it? Suppose there is, and that 'large' objects **M** and **N** are somewhere above that level. Then $m\langle \mathbf{M}, \mathbf{N}\rangle$ *cannot* contain more information than $m(\mathbf{M})$ and $m(\mathbf{N})$ combined, so that $m\langle \mathbf{M}, \mathbf{N}\rangle \Leftrightarrow m(\mathbf{M})\ \&\ m(\mathbf{N})$. **M** and **N** cannot be non-classically correlated with each other, no matter how much they interact with each other. In this case the description $m(\mathbf{M})$, say, is not just maximal in the usual sense that there is no better description of the system **M** considered by itself. It is also maximal in some stronger sense. We might say that the model $m(\mathbf{M})$ is *saturated*.

What is this 'stronger sense' of maximal? Let us consider two systems **X** and **Y** that are non-classically correlated, so that $m\langle \mathbf{X}, \mathbf{Y}\rangle \Rightarrow m(\mathbf{X})\ \&\ m(\mathbf{Y})$. This means that $m\langle \mathbf{X}, \mathbf{Y}\rangle$ says more about the systems **X** and **Y** than do $m(\mathbf{X})$ and $m(\mathbf{Y})$ together. Now, ontologically speaking, the combined system $\langle \mathbf{X}, \mathbf{Y}\rangle$ is nothing over and above the two systems **X** and **Y**. Since $m(\mathbf{X})$ and $m(\mathbf{Y})$ are maximal descriptions of **X** and **Y**, then, what is the extra information in $m\langle \mathbf{X}, \mathbf{Y}\rangle$ supposed to be *about*? It cannot be about **X**, as $m(\mathbf{X})$ is maximal concerning **X**, nor about **Y**, since $m(\mathbf{Y})$ is maximal concerning **Y**; but then there is nothing left! There is clearly some sense in which $m(\mathbf{X})$ and $m(\mathbf{Y})$ are not maximal after all. The extra information concerns the relation between **X** and **Y**, which is indirectly information about both **X** and **Y**. The descriptions $m(\mathbf{X})$ and $m(\mathbf{Y})$ are only the best descriptions of **X** and **Y** *considered in isolation*; they are not maximal regarding these systems *tout court*. A saturated description $m(\mathbf{M})$, by contrast, is maximal *tout court*. It not only provides maximal information about **M** considered as a single system but also, in conjunction with $m(\mathbf{N})$, provides maximal information about the *relation* between **M** and **N**, as long as $m(\mathbf{N})$ is also saturated.

Do such systems with saturated models actually exist, however? If they do, then this gives us a promising candidate for what is fundamentally different about classical and quantum systems. The classical world, containing the 'macroscopic' objects of ordinary experience, such as chairs, beetles, automobiles, and measuring devices may just be the world above the level of saturation.[25] From this level up, there are no more logically emergent properties. This would help to explain why logical emergence is so counter-intuitive. I have no proof that there are objects with saturated models, nor much idea of where the line between classical and quantum objects may be. In any case, the line is surely quite fuzzy, as it is the sort of thing that is approached asymptotically. It may even lie at a higher scale for complex, biological systems than for simple, inanimate objects. We shall assume that a level of saturation exists, however, to see if it is of any help in the interpretation of QM.

In the remainder of this chapter the terms 'classical,' 'quantum,' 'large,' 'small,' 'macroscopic,' and 'microscopic' must be used with great caution, in order to avoid tautologies and circular arguments. Though it is undesirable to redefine words that are in common use, I am forced to make the following stipulative definitions. Fortunately, the meanings given are within the range of standard usage.

**7.2.2.1 Definitions** (1) A *classical* system is one that can be described by classical physics.

(2) A *quantum* system is one whose states obey the superposition principle.

(3) A *macroscopic* system is one that is visible to the naked eye.

(4) A *microscopic* system is one that is roughly the size of an atom.

(5) A *small* system is one that does not have a saturated model.

(6) A *large* system is one that has a saturated model.

It should be noted that there is a huge gap between macroscopic and microscopic systems, as defined here. This is quite deliberate. It is common practice, of course, to apply quantum mechanics to microscopic systems and classical physics to macroscopic systems. I hold therefore that there is at least a rough equivalence between quantum and microscopic systems, and between classical and macroscopic systems. My more original contention, however, is that there is a similar correspondence between small and quantum systems, and between large and classical systems.

### *7.2.3 Relative Models for Small Systems*

In chapter 6 it was argued that the failure of a maximal description $m\langle \mathbf{X}, \mathbf{Y}\rangle$ to factorize is due to a special, *sui generis* physical relation between the systems **X** and **Y**. Since the description is maximal, the failure to factorize cannot be *merely* a matter of knowledge, as it is in the coffee mug example, but rather a reflection of the physical situation. An example of such a physical relation was that which holds between the two fragments of a sheet of paper after it has been torn in half. If two systems are related in such a manner as this, they must have interacted in the past, since separate causal processes always give rise to separate histories.

Let us consider two systems, **X** and **Y**, that interacted in the past but are now isolated. In general they will be correlated, so that $m\langle \mathbf{X}, \mathbf{Y}\rangle \Rightarrow m(\mathbf{X})$ & $m(\mathbf{Y})$, and we cannot factorize $m\langle \mathbf{X}, \mathbf{Y}\rangle$ into separate absolute descriptions of **X** and **Y**. We may still factorize $m\langle \mathbf{X}, \mathbf{Y}\rangle$, however, using relative propositions. Recall from Theorem 2.2.4.4 that, if $A \Rightarrow B$, we may express $A$ as the conjunction ($B$ & $A/B$), where $A/B$ is the relative state of affairs $A$ given $B$, as defined in Definition 2.2.4.1. In this case, since $m\langle \mathbf{X}, \mathbf{Y}\rangle \Rightarrow m(\mathbf{Y})$, we get the following equivalence.

$$m\langle \mathbf{X}, \mathbf{Y}\rangle \Leftrightarrow m(\mathbf{Y})\ \&\ m\langle \mathbf{X}, \mathbf{Y}\rangle/m(\mathbf{Y}).$$

The symbol '$m\langle \mathbf{X}, \mathbf{Y}\rangle/m(\mathbf{Y})$' denotes everything $m\langle \mathbf{X}, \mathbf{Y}\rangle$ has to say that is not already said by $m(\mathbf{Y})$. For convenience, we will write $m\langle \mathbf{X}, \mathbf{Y}\rangle/m(\mathbf{Y})$ in the abbreviated form '$m(\mathbf{X}, \mathbf{Y}/\mathbf{Y})$,' and call it the maximal proposition concerning $\langle \mathbf{X}, \mathbf{Y}\rangle$ relative to **Y**. It is clear that $m(\mathbf{X}, \mathbf{Y}/\mathbf{Y}) \Rightarrow m(\mathbf{X})$, but not vice versa. We could, of course, exchange **X** with **Y** in these relations, and write: $m\langle \mathbf{X}, \mathbf{Y}\rangle \Leftrightarrow m(\mathbf{X})$ & $m(\mathbf{X}, \mathbf{Y}/\mathbf{X})$.

Now suppose that **X** is a small system whereas **M** is large, in the sense of Definition 7.2.2.1, and that we want to describe **X** as completely and in as much detail as possible. Since **M** is large, it follows that $m(\mathbf{M})$ is not merely maximal but also saturated. We then have

$$m\langle \mathbf{X}, \mathbf{M}\rangle \Leftrightarrow m(\mathbf{M})\ \&\ m(\mathbf{X}, \mathbf{M}/\mathbf{M}).$$

Since $m(\mathbf{M})$ is saturated, and there is no overlap of content between $m(\mathbf{M})$ and $m(\mathbf{X}, \mathbf{M}/\mathbf{M})$, it follows that $m(\mathbf{X}, \mathbf{M}/\mathbf{M})$ tells us nothing about **M** at all. It says nothing about **M** considered as an individual system, nor about how it is related to $\mathbf{X}$. From this it follows that $m(\mathbf{X}, \mathbf{M}/\mathbf{M})$ is entirely about the system $\mathbf{X}$! This is perhaps surprising, as $m(\mathbf{X}, \mathbf{M}/\mathbf{M})$ is stronger than $m(\mathbf{X})$, which is maximal concerning $\mathbf{X}$ considered as an individual. $m(\mathbf{X}, \mathbf{M}/\mathbf{M})$ goes beyond $m(\mathbf{X})$ by describing $\mathbf{X}$'s relation to $\mathbf{M}$ – something that $m(\mathbf{X})$ does not do. Since $m(\mathbf{X}, \mathbf{M}/\mathbf{M})$ is a description of $\mathbf{X}$ only, and says nothing about $\mathbf{M}$, we shall henceforth write it as $m(\mathbf{X}/\mathbf{M})$.

We see that, unless perhaps $\mathbf{X}$ is also correlated with other systems, the best available description of $\mathbf{X}$ is not the absolute description $m(\mathbf{X})$, but the *relative* description $m(\mathbf{X}/\mathbf{M})$.[26] This is the best description of $\mathbf{X}$ alone, and is made from the vantage point of someone having saturated knowledge about $\mathbf{M}$, to which $\mathbf{X}$ is physically related. I do not mean to suggest that $m(\mathbf{X})$ gives no information at all about $\mathbf{X}$, however. If $\mathbf{X}$ is a alpha particle, for example, then $m(\mathbf{X})$ will at least specify the mass and charge of the system, as these can be described even in the absence of any correlation with a large system.

The above discussion assumes that $\mathbf{X}$ is correlated to the whole of $\mathbf{M}$, not merely to a part, or a number of separate parts, of $\mathbf{M}$. What does this mean? Suppose $\mathbf{M}$ can be mentally divided into two parts $\mathbf{M}_1$ and $\mathbf{M}_2$, which may not themselves be large. If $\mathbf{X}$ is correlated with $\mathbf{M}_1$ alone, and not the whole of $\mathbf{M}$, then we have

$$
\begin{aligned}
m\langle\mathbf{X}, \mathbf{M}\rangle \Leftrightarrow m\langle\mathbf{X}, \mathbf{M}_1, \mathbf{M}_2\rangle &\Leftrightarrow m\langle\mathbf{X}, \mathbf{M}_1\rangle \,\&\, m\langle\mathbf{M}_1, \mathbf{M}_2\rangle \\
&\Leftrightarrow m(\mathbf{M}_1) \,\&\, m(\mathbf{X}, \mathbf{M}_1/\mathbf{M}_1) \,\&\, m\langle\mathbf{M}_1, \mathbf{M}_2\rangle \\
&\Leftrightarrow m(\mathbf{M}) \,\&\, m(\mathbf{X}, \mathbf{M}_1/\mathbf{M}_1).
\end{aligned}
$$

Note that the term $m\langle\mathbf{M}_1,\mathbf{M}_2\rangle$ cannot be factorized, since $m(\mathbf{M}_1)$ and $m(\mathbf{M}_2)$ may not be saturated, so that $\mathbf{M}_1$ and $\mathbf{M}_2$ may be correlated. In this case the relative model $m(\mathbf{X}, \mathbf{M}_1/\mathbf{M}_1)$ may add to the information about $\mathbf{M}_1$ in $m(\mathbf{M}_1)$, so that it cannot be considered a description of $\mathbf{X}$ alone. Thus, if $\mathbf{X}$ is correlated only to $\mathbf{M}_1$ and not to the whole of $\mathbf{M}$, $m(\mathbf{X}, \mathbf{M}/\mathbf{M})$ cannot be replaced by $m(\mathbf{X}/\mathbf{M})$, and may not be a relative model of $\mathbf{X}$. A similar situation obtains if $\mathbf{X}$ is *separately* correlated to both $\mathbf{M}_1$ and $\mathbf{M}_2$, for this gives us

$$
\begin{aligned}
m\langle\mathbf{X}, \mathbf{M}\rangle &\Leftrightarrow m\langle\mathbf{X}, \mathbf{M}_1, \mathbf{M}_2\rangle \\
&\Leftrightarrow m\langle\mathbf{X}, \mathbf{M}_1\rangle \,\&\, m\langle\mathbf{X}, \mathbf{M}_2\rangle \,\&\, m\langle\mathbf{M}_1, \mathbf{M}_2\rangle \\
&\Leftrightarrow m(\mathbf{M}_1) \,\&\, m(\mathbf{X}, \mathbf{M}_1/\mathbf{M}_1) \,\&\, m(\mathbf{M}_2) \,\&\, m(\mathbf{X}, \mathbf{M}_2/\mathbf{M}_2) \,\&\, \\
&\qquad m\langle\mathbf{M}_1, \mathbf{M}_2\rangle \\
&\Leftrightarrow m(\mathbf{M}) \,\&\, m(\mathbf{X}, \mathbf{M}_1/\mathbf{M}_1) \,\&\, m(\mathbf{X}, \mathbf{M}_2/\mathbf{M}_2),
\end{aligned}
$$

and we again have no relative model of $\mathbf{X}$ alone.

So, for $m(\mathbf{X}/\mathbf{M})$ to exist, as a description of **X** alone, **X** must be correlated with **M** as a whole. This would seem to require that **X** have interacted with **M** as a whole, and not just with some part of **M**, or separately with two or more parts of **M**.

The notion of a relative model, defined as a relative state of affairs in the sense of Definition 2.2.4.1, is essential to the understanding of quantum mechanics being developed here. The idea that quantum-mechanical descriptions are relative descriptions is not new, of course. Niels Bohr, for example, found strong parallels between the role of measuring instruments in quantum mechanics and reference systems in relativity theory. He writes, for instance:

> I should still like to emphasize the bearing of the great lesson derived from general relativity theory upon the question of physical reality in the field of quantum theory. In fact, notwithstanding all characteristic differences, the situations we are concerned with in these generalizations of classical theory present striking analogies which have often been noted ... The dependence on the reference system, in relativity theory, of all readings of scales and clocks may even be compared with the essentially uncontrollable exchange of momentum or energy between the objects of measurements and all instruments defining the space-time system of reference, which in quantum theory confronts us with the situation characterized by the notion of complementarity. In fact, this new feature of natural philosophy means a radical revision of our attitude as regards physical reality, which may be paralleled with the fundamental modification of all ideas regarding the absolute character of physical phenomena, brought about by the general theory of relativity. (1935: 150–1)

More recently, Everett (1957) has drawn attention to the apparent relativity of quantum states, in his proposal that the usual state vector in fact represents the *relative* state of a system, relative to some sub-space of the Hilbert space of an observer system. Everett writes:

> Thus we are faced with a fundamental *relativity* of states, which is implied by the formalism of composite systems. It is meaningless to ask the absolute state of a subsystem – one can only ask the state relative to a given state of the remainder of the subsystem. (1957: 317)

This relativistic aspect of Everett's ideas has been somewhat ignored, with most critics focusing on the 'many worlds' version of the theory, but it is restored to its proper place in some recent followers of Everett's approach, such as Saunders (1995) and Lockwood (1989: chap. 13).

It is unfortunate that Einstein gave his theories the name 'relativity,' as there is nothing particularly relativistic about them.[27] They do involve relativistic notions, such as the idea that some physical quantities previously supposed to be absolute, such as durations and lengths, turn out to be defined only relative to a reference frame. However, the scope of this relativism is somewhat limited, since all quantities are definable with the invariant structure of Minkowsky spacetime. In relativity theory, in other words, relative quantities (such as durations) are *reducible* to relations within an absolute description, involving spacetime intervals, four-vectors, and so on.

In quantum theory, by contrast, the relativity of some physical quantities is *irreducible*. Quantum mechanics, we might say, is the true theory of relativity. QM deals in states that are essentially relative, since there is no mathematical description of the quantum world that is absolute. In QM there is no God's-eye view, no counterpart of Minkowsky spacetime.

It is also worth noting that the definition of relative models given here is based upon a new logical concept – that of a relative state of affairs. An understanding of relative models thus requires the adoption of a new logic. This relativity logic is not revisionary, however, as it does not challenge any of the hallowed truths of classical logic. It is merely a supplement to the standard logic of absolute states of affairs, and is in conformity with common sense. Moreover its original introduction, in §2.2.4, was motivated quite independently of quantum mechanics.[28]

### *7.2.4 Measuring Instruments*

On the view being developed here, pieces of laboratory apparatus, including devices for preparing and measuring quantum systems, are large systems. We shall now investigate whether this idea helps to make sense of quantum measurement.

A measuring device **M** must have some attribute that serves to indicate the outcome of each measurement, such the position of a needle or the pattern shown on an LED display. We will represent this 'pointer' variable as $Q$. Before a given measurement the apparatus is set to some standard initial 'zero' state. The measurement process itself is an interaction between **M** and the system **X**, which leaves the pointer in some final position $Q = q$, so that $q$ is the outcome of the measurement. It is well known that, for a quantum system **X**, we cannot in general regard a measurement as revealing some pre-existing attribute of the system. Many measurement processes are stochastic, in the sense that the final value of $Q$ cannot be predicted in advance, even from maximal information. Thus, we cannot say that **M** is a proper measuring device just in case it reliably produces the right values of $Q$ for the right systems. So what is it that makes **M** a measurement apparatus, rather than some foolish contraption?

The standard answer to this is as follows. Suppose the measurement apparatus measures some quantity $A$ on the system, whose possible values are the eigenvalues $a_i$ of $A$, and whose corresponding eigenvectors are $|a_i\rangle$. There is then a bijection between the eigenvalues $a_i$ and the possible pointer readings $q_i$, so that if the system is measured when initially in state $|a_i\rangle$, then the pointer necessarily ends up at $q_i$. To put it briefly, if $A$ has a definite value, then a measurement using $Q$ is guaranteed to reveal that pre-existing value. In this case, to call the experiment a 'measurement' seems appropriate.

What if the system's pre-measurement state is not an eigenstate of $A$, however? What will happen when an experiment to measure $A$ is performed? If the initial state of the system is $|\psi\rangle$, then the outcome of the measurement could be any one of the positions $q_i$ for which the projection $\langle a_i|\psi\rangle$ is non-zero, as each outcome $q_i$ has chance $|\langle a_i|\psi\rangle|^2$. In this case, however, it is hard to see why the outcome $Q = q_i$ should be called a measurement result, as it does not reflect the prior state of the system measured. All we can say is that the quantum system caused the pointer $Q$ to adopt the final position $q_i$.

In one kind of measurement, called a measurement of the first kind, the outcome $Q = q_i$ does reflect the state of the measured system, but only *after* the measurement has occurred. After such a measurement with outcome $Q = q_i$, the system is left in the eigenstate of $A$ corresponding to this outcome, namely $|a_i\rangle$. This type of experiment is also called a *preparatory* measurement, since it leaves the system in a prepared state. One important feature of a preparatory measurement is that a second measurement performed on the same system, immediately after the first, will always yield the same outcome as the first.[29] Indeed, in some philosophical discussions of quantum measurement, this consistency of outcomes in repeated experiments is held up as the defining characteristic of a measurement.[30]

Unfortunately, although preparatory measurements are the kind most often discussed in textbooks, the majority of actual laboratory experiments are not of this kind. In general, the measured system is not left in *any* eigenstate of the observable measured, let alone the eigenstate whose corresponding eigenvalue is the measurement result. Indeed, in many cases the system measured no longer even exists after the measurement is complete, as for example when a photon is detected by a photographic plate. In this kind of measurement, called a measurement of the second kind, the outcome of the experiment $Q = q_i$ does not (in general) represent the state of the system at any time.

We see that Bell (1990: 34) is right to say that the term 'measurement' is misleading in quantum mechanics, so that the word 'experiment' might be preferable. Even in a preparatory measurement, the outcome corresponds to the state of the system only after the interaction, in general; in a measurement of the second kind the 'measured value' seems to have no significance at all, in relation to the system measured. This should be borne in mind during the following discussion.

Of the two kinds of measurement, it is preparatory (or 'ideal') measurements that are the more interesting. It is here that the measurement process seems to be endowed with a special physical significance, since it is the occasion of the infamous 'collapse' of the wavefunction into an eigenstate of the observable. The following discussion of the measurement process focuses therefore on ideal measurements.

As stated above, the first ideal measurement of a given type performed on a system usually has a random outcome, but a second measurement of the same type, made immediately after the first, is usually guaranteed to have the same outcome as the first. Thus, while the first process is stochastic, the second is deterministic. Suppose measurements are made by **M** on **X** at $t_1$ and $t_2$, both yielding $Q = q$. What information can be used to predict the outcome of the second measurement? It must be maximal information about the two systems involved in the process, that is, $m$(**M**) & $m$(**X**/**M**),[31] just before $t_2$. Thus, if the first measurement has outcome $Q = q$, say, then the outcome $Q = q$ for the second measurement may be inferred from $m$(**M**) & $m_t$(**X**/**M**), for $t_1 < t < t_2$. The description $m$(**M**) can be regarded as independent of the measurement outcome, as **M** is reset after each measurement, so the result $Q = q$ must be somehow contained in the relative model $m_t$(**X**/**M**).

It follows that, if **M** is an ideal measuring device, then after each measurement is finished **M** displays information about the relation between **M** and **X**, indicated by its pointer. The pointer does not indicate any *pre*-existing feature of the system, but rather a feature that exists *after* the measurement is complete. Moreover, although this final feature exists within the system **X**, ontologically speaking, it can only be described relative to the measurement apparatus. More precisely, it is the system **X** that acts on **M** to *cause* the second outcome $Q = q$, yet this result cannot be *predicted* from $m$(**X**) alone – the more detailed relative description $m$(**X**/**M**) is required.

Suppose we have two measuring instruments **M** and **N**, of different types, and measure **M**, then **N**, then **M** all on the same system **X**, at times $t = 1, 2, 3$. Let $m_t$(**X**/**M**) denote the model of **X** relative to **M** just *after* time $t$. For some devices **M** and **N**, although not others, the result of the second measurement by **M** is not guaranteed to equal the first. Indeed, in such cases the second measurement by **M** has a random outcome, so that the result cannot be inferred from $m_2$(**X**/**M**) & $m_2$(**X**/**N**). Now this is a little odd since, as argued above, the measurement outcome for **M** is contained in the relative model $m_1$(**X**/**M**). Thus, if $m_2$(**X**/**M**) were equal to $m_1$(**X**/**M**), the measurement outcome would also be entailed by $m_2$(**X**/**M**) & $m_2$(**X**/**N**). (The mere adding of extra information, namely $m_2$(**X**/**N**), should not prevent the value of $Q$ from being inferred from $m_2$(**X**/**M**)! We are thus forced to conclude that $m_2$(**X**/**M**) is quite distinct from $m_1$(**X**/**M**). Somehow the correlation between **X** and **M**, forged during their interaction, is destroyed or messed up by the measurement of **X** by **N**.

In view of my theory that a non-classical correlation between two systems depends on their having some unique similarity, the following analogy suggests itself here. Imagine a wax tablet **X**, into which a seal **M** is pressed.[32] The tablet now bears an exact imprint of the seal, so that the two are similar in a very special respect. (Indeed, it is this very fact that the shape of the wax requires a previous interaction with the seal that gave a document thus marked its authority.) Suppose now that a second seal, **N**, with a different pattern, is pressed into the same tablet **X**. This results not only in **X** becoming similar to **N**, but also in **X** *ceasing* to be similar to **M**. Provided the wax is reasonably soft and malleable, the previous imprint of **M** is simply obliterated by the new imprint of **N**.

It is important to realize that this situation, where a measurement by **N** seems to erase the result of an earlier ideal measurement by **M**, does not exist for every pair of devices **M** and **N**. This erasure only occurs when the instruments are represented, in QM, by operators that do not commute. For a pair of instruments whose operators commute, it seems that the system can be correlated to both of them at once. The possibility of **X**'s simultaneous correlation with **M** and **N** depends on the concrete natures of the three bodies concerned – it has nothing to do with description. I do not pretend to have any insight into why some pairs of devices are compatible, and others incompatible, but a second analogy may be of some help.

Consider a rectangular plastic lunch box, with a lid that snaps onto the rim. Sometimes it is hard to find the right lid to a particular box, and on occasion I have tried to use a lid that is slightly too small. What then happens is that it is impossible to fit the lid completely onto the box. It is easy enough to get two diagonally opposite corners to snap on at one time, but then the other two will not go on. If you try to force the second two corners on, then the first two always slip off. The fitting of each corner, in other words, in physically incompatible with the fitting of either corner adjacent to it, but compatible with the one diagonally opposite. I have no doubt that, with a little work, an applied mathematician could figure out exactly why these relations obtain, in terms of the different forces required to stretch the lid in different directions, and its tendency to shrink in one dimension when stretched in another.

In a similar way, there are no doubt good physical reasons why some pairs of instruments can be physically correlated with a system at the same time, whereas others cannot. The difference is that in the quantum case we cannot achieve the same level of understanding of why the correlations for a given pair of devices exclude one another. These correlations are the very means by which we are able to describe a small system in any detail at all, so we seem to be unable, even in principle, to describe the correlations themselves.

The wax tablet analogy has another virtue that deserves mention. When the tablet interacts with the seal, the result is that the two systems become similar, but we should note that only the tablet is physically affected to any significant degree. The tablet takes on the pattern of the seal, not the other way around, which is an important asymmetry. This is quite different from the correlation between the two halves of the torn sheet of paper, which is symmetric. In a measurement of a microscopic system it seems that, as in the wax tablet example, the apparatus leaves its imprint on the system. If there is another device **M′** exactly similar to **M** then, after a measurement of **X** by **M**, **X** is correlated just as much with **M′** as with **M**, even though **M′** has not interacted with anything. Although **M** is affected by the measurement interaction, as its pointer must move, this change is irrelevant to the correlation between **M** and **X**. (The correlation remains, for instance, when **M** is reset to its standard pre-measurement state.) We might say that the **M–X** correlation is *one way*, as it consists of just one system, **X**, changing to match **M**.

The main drawback of the wax tablet picture it that it allows only one type of correlation between the tablet and a given seal. In the case of a measuring instrument and a small system, however, there are always at least two possible types of correlation, and often many more than that. Roughly speaking, each possible outcome of the measurement interaction corresponds to a different manner of correlation between the two systems. The measurement apparatus does not therefore force a single type of relation between the systems, but ‘gives the system a choice’ from a list of possible relations, which are represented in QM by the eigenstates of the observable.

### *7.2.5 States for Small Systems*

We learn from QM that observables can be grouped together into ‘complete sets,’ which are such that (1) any two observables in the set commute, and (2) every observable not in the set fails to commute with any of those in the set. The operators of a complete set seem then to correspond to a set of measuring devices that can all be simultaneously correlated with a given system, without getting in each other's way. Suppose that $\{\mathbf{M}_i\}$ is such a complete set of $n$ compatible devices, and that each makes a measurement of the system **X**, either all simultaneously or in quick succession. The best description of the **X** is now given by the conjunction of relative models $m(\mathbf{X}/\mathbf{M}_1)$ & $m(\mathbf{X}/\mathbf{M}_2)$ & ... & $m(\mathbf{X}/\mathbf{M}_n)$, which we can abbreviate to '$\&m(\mathbf{X}/\mathbf{M}_i)$.'

This description $\&m(\mathbf{X}/\mathbf{M}_i)$ is maximal in two different senses. First, it is the best description of **X** that exists in **X**'s current situation. Second, **X**'s current situation is also the best possible one, as far as facilitating descriptions of **X** is concerned. As an analogy, consider a hockey team whose roster is weakened by a number of injuries to key players. In that situation, suppose it plays as well as it possibly can, given the skaters available. This performance is maximal given the circumstances, but the circumstances themselves are less than ideal. The team could have played better with a deeper line-up. Now suppose that, later in the season, the roster is completed as the players return from injury, and the team once again plays as well as it possibly can. In this case the performance is maximal in two senses; it is the best performance from the best squad. Short of changing the team itself by making trades, there is no way to improve on the performance. In a similar way, the proposition $\&m(\mathbf{X}/\mathbf{M}_i)$ is not only the best description of **X** under the present circumstances, but is at least as detailed as any best description would be under different circumstances. We shall therefore say that the description $\&m(\mathbf{X}/\mathbf{M}_i)$, which consists of maximal relative models for a complete set of measuring instruments, is *saturated.* A saturated relative model will be called a *correlation state*, or just *state*.

The fact that there are such things as complete sets of measuring instruments suggests that a small system has a kind of correlation *capacity* that cannot be exceeded. Once a small system is fully correlated, new correlations may be formed only at the expense of some of the old ones. Also, judging from the structure of QM, it seems that the correlation capacity of a small system may be filled equally by other small systems as by large systems. If **X** is correlated with another small system **Y**, however, this does not provide any means for a description of **X**, as $m(\mathbf{X}, \mathbf{Y}/\mathbf{Y})$ concerns **Y** almost as much as it concerns **X**. Thus, the correlation with **Y** robs **X** of a saturated relative model, or state.

## 7.3 CHANCE FOR SMALL SYSTEMS

Now that we have examined the way in which information about small systems is represented, we are ready to discuss the chances of events in a small system, which we may call 'small events.' A small event can only be described, in any detail, if the small system is correlated with a large one, such as a measuring instrument, with a saturated model. Thus, the chance of a small event is always the chance of some relative state of affairs *m*(**X**/**M**). In an ideal measurement, each final relative state of affairs *m*(**X**/**M**) for the small system is uniquely associated with some macroscopic state of the apparatus, so that the small event has the same physical chance as the corresponding measurement outcome.

The up-to-date (utd) chance function $P_t$ seems to be the more useful version of the chance function here, which is based on maximal knowledge of the dynamics of the system, the boundary condition and the history of the world up to time *t*. To define the chance function for a system **X**, one might think that it is sufficient to describe the past history of **X** alone, so that, for some microscopic event *A* in **X**, $P_t(A) = P_r(A|\mathcal{L}_{\mathbf{X}}\ \&\ m_t(\mathbf{X}/\mathbf{M}))$.[33] This approach is only valid, however, when **X** is not correlated with any other small system, as we shall now see.

### *7.3.1 The Nonlocality of Chance*

Let us consider again the EPR experiment, in which small systems **X** and **Y** are initially correlated with each other. Since they are correlated with each other, neither is individually correlated with a complete set of measuring instruments, although we assume that the pair ⟨**X**, **Y**⟩ has a state. In other words, the relative model *m*(⟨**X**, **Y**⟩/**M**) is saturated, but it does not factorize into *m*(**X**/**M**) & *m*(**Y**/**M**), since these relative models *m*(**X**/**M**) and *m*(**Y**/**M**) are far from saturated. We then individually measure the systems **X** and **Y**, using some apparatus **N** that has two possible outcomes for each experiment. Let us suppose that from *m*(⟨**X**, **Y**⟩/**M**) we can infer with certainty that the two results will be opposite, even though we cannot predict the outcome of either experiment.

If we measure **X** first, then **Y**, then the result of **Y**'s measurement is predictable in advance, with certainty, using the outcome of **X**'s measurement. Thus, the chance function for **Y** is altered by the result of **X**'s measurement – an outcome of spin up on **Y** might increase from $\frac{1}{2}$ to 1, perhaps. This is so even if the systems are far apart when **X** is measured, so that the chance function for **Y** may depend on events which occur far away. At exactly *what time*, at the system **Y**, does the chance increase from $\frac{1}{2}$ to 1? One would like to say that the change occurs when **X** is measured, but this is ruled out by special relativity. There is no time at **Y** that is simultaneous with the measurement on **X**, and so 'the time of **X**'s measurement' does not pick out any unique time at **Y**. The chance function, though defined for events in spacetime, does not itself exist in spacetime.

When **X** is correlated with other small systems, therefore, we cannot define the chance function for **X** by reference to **X** alone. We have to consider a larger system, namely, the smallest system that (1) includes **X** and (2) is fully correlated to large objects, so that its relative model is saturated. The chance function thus obtained, for ⟨**X**, **Y**⟩ in this case, is obviously a joint chance for **X** and **Y** together, which cannot be factorized in separate distributions for **X** and **Y**, although **X** and **Y** do of course have their own marginal distributions.

### *7.3.2 Coordinates for Chance*

Consider a set of measuring instruments $\{\mathbf{M}_i\}$, each of which is able to make a complete measurement on a small system **X**. The possible states for **X** are then of the form $\mathbf{M}_r^a$, that is, a correlation of type *a* for the *r*th instrument, if we assume for now that **X** is not correlated to other small systems. For example, if **X** is a spin-$\frac{1}{2}$ particle whose motion can be ignored, then each $\mathbf{M}_r$ is a device for measuring the intrinsic spin of the system along some particular axis, and *a* is one of two values, for spins up and down along that axis. Chances attach to *transitions* between pairs of such states, such as from $\mathbf{M}_4^+$ to $\mathbf{M}_6^-$, where '+' and '−' denote spin up and spin down.

It would be very convenient if there were some way to represent all the possible correlations, that is, all the states $\mathbf{M}_r^a$, on the same map, so to speak. As it is, since each represents a correlation with a differently aligned device, the different possible states are completely unrelated. Essentially, each device $\mathbf{M}_r$ defines a *separate* configuration space of (two) possible states for **X**, which we would like somehow to bring together, perhaps by embedding them all in a single, larger space. This unified structure would have to be based on a deeper representation of the system, as it must see relations between the different types of correlation, rather than viewing them as completely different kinds of animal. This deeper representation might even enable one to derive the chance of a given transition between two correlations from the relation between these states.

Putting it another way, we would like there to be a kind of *coordinate system* for physical chance, analogous to the coordinate systems in geometry that generate distances. A coordinate system in geometry assigns an ordered triple of numbers to each particle in a rigid system, thus putting all the particles together on the same map, so to speak. Moreover, the distance between any two particles (i.e., the *d*-list, as defined in §7.1.1) can be generated from the coordinates of those particles, using the Euclidean metric. In a similar way, we would like there to be a set of 'coordinates' for each correlation state $\mathbf{M}_r^a$, which enables these states to be put onto the same map. As a bonus, the chance of a transition from $\mathbf{M}_r^{\,a}$ to $\mathbf{M}_s^{\,b}$ might even be derivable as a function of the coordinate sets of these two states, that is, using a metric on the unified state space.

I see no a priori reason to hope that such a 'coordinate structure' exists, but in fact just such a representation is provided by quantum mechanics. Suppose $\mathbf{M}_0$, for instance, represents a device for measuring spins along the positive $z$-axis. Then $\mathbf{M}_0$ assigns a 'coordinate pair,' that is, a wavefunction, $(w, z)$ to each possible spin state of the system $\mathbf{M}_r^{\,a}$, where $w$ and $z$ are complex, and normalized so that $|w|^2 + |z|^2 = 1$. This coordinate pair $(w, z)$ may be calculated from the angle, relative to the $z$-axis, along which the instrument $\mathbf{M}_r$ measures spins, using some spherical trigonometry.[34] Suppose $\mathbf{X}$ has the spin state $(w, z)$, and we make a measurement of it using $\mathbf{M}_0$. Then, according to QM, the chance of measuring spin up is $|w|^2$, and the chance of spin down is $|z|^2$. In this special case where the device making the measurement is also the one that defines the 'coordinate system,' the squared 'coordinates' are just physical chances.

In the more general case, where a system in state $(w, z)$ is measured by some other apparatus $\mathbf{M}_r$, we need to use the vector-space formalism. We define a state vector $|\psi\rangle$ as a function from wavefunctions to wavefunctions, which maps $(w, z)$ to $(w + a, z + b)$, where $a$ and $b$ are also complex. Each spin state vector is therefore characterized by a pair of complex numbers $a$ and $b$, which may be written as a column vector. This state vector maps the origin $(0, 0)$ to the wavefunction $(a, b)$, and so is the position vector of $(a, b)$. In general, the position vector of a spin wavefunction $\psi$ is written $|\psi\rangle$, and is known as a *spinor*. We can define the scalar product of two vectors as in §6.1.2, so that

$$\begin{pmatrix} a_1 \\ a_2 \end{pmatrix} \begin{pmatrix} b_1 \\ b_2 \end{pmatrix} = a_1^* b_1 + a_2^* b_2 \,,$$

and say that they are *orthogonal* just in case the scalar product is zero. If we add the usual operation of multiplication of a vector by a (complex) scalar, then we have a vector space $H$. We can now forget about the original wavefunction structure, as everything of importance is included in the vector-space structure.

Each measuring instrument $\mathbf{M}_r$ defines an orthogonal basis on the vector space $H$, as follows. If the wavefunctions $r_1$ and $r_2$ correspond to the states $\mathbf{M}_r^{\,+}$ and $\mathbf{M}_r^{\,-}$ respectively, then it turns out that the position vectors $|r_1\rangle$ and $|r_2\rangle$ are orthogonal. We also normalize these vectors, so that $|r_1\rangle.|r_1\rangle$ and $|r_2\rangle.|r_2\rangle$ each equal unity. (The state $\mathbf{M}_0^{\,+}$ has wavefunction $(1, 0)$, and $\mathbf{M}_0^{\,-}$ has $(0, 1)$, for example, and the position vectors for these are clearly orthogonal.) The vector-space structure therefore allows us to represent the spin state of a system with respect to any apparatus $\mathbf{M}_r$ rather than just with respect to $\mathbf{M}_0$. Instead of writing a state as a wavefunction $\psi = (w, z)$, we may write it as a vector $|\psi\rangle$, which can be expressed using any set of base vectors such as $|r_1\rangle$ and $|r_2\rangle$, that is, $|\psi\rangle = c_1|r_1\rangle + c_2|r_2\rangle$. If we had chosen some other apparatus $\mathbf{M}_j$ instead of $\mathbf{M}_0$ as the reference system for the wavefunctions, then the wavefunction $(w, z)$ for each state would have been different. Values like $c_1$ and $c_2$, by contrast, are independent of the reference system for the wavefunctions.

Now that we can represent spin states relative to a general apparatus $\mathbf{M}_r$, we can give the chances of measurement outcomes for $\mathbf{M}_r$. If the system is originally in state $\psi = (w, z)$, we express this in the orthonormal basis for $\mathbf{M}_r$, to get $|\psi\rangle = c_1|r_1\rangle + c_2|r_2\rangle$. The chance of spin up is then $|c_1|^2$, and the chance of spin down is $|c_2|^2$. Since a spin-up result leaves the system in the state $|r_1\rangle$, and a spin-down result leaves it in $|r_2\rangle$, the chance of a general transition from $|\psi\rangle$ to $|\phi\rangle$, upon measurement of some operator for which $|\phi\rangle$ is an eigenstate, where both vectors are normalized, is $||\phi\rangle.|\psi\rangle|^2$, which in the Dirac notation introduced in §6.1.2 is written $|\langle\phi \mid \psi\rangle|^2$. It is easy to show that $c_i = \langle r_i \mid \psi\rangle$.

We see that there is an analogy, at least, between coordinate sets and wavefunctions. The 'points' to which wavefunctions are assigned are correlation states, and the metric defined on pairs of such points, that is, the squared modulus of the overlap integral, represents physical chance. In geometry it is more convenient to use the vector-space formalism rather than the coordinate formalism, as this representation has no privileged origin, and no preferred directions (i.e., coordinate axes) either. Instead of giving the position of a point as $(x, y)$, we specify its position relative to some other point $P$ as $a\mathbf{i} + b\mathbf{j}$, where $\{\mathbf{i}, \mathbf{j}\}$ is some arbitrary basis of the vector space. In a similar way the vector-space formalism is also preferable in QM, as it enables one to treat all observables on an equal basis, so that one deals with state vectors rather than with wavefunctions.

The analogy between wavefunctions and coordinate sets is my main heuristic in searching for an interpretation of the wavefunction, or state vector. In the next section, therefore, I will look more deeply into the nature of geometrical coordinates, in order to be able to develop the analogy further.

### *7.3.3 The Nature of Coordinates*

There is no disagreement about the empirical meaning of rectangular cartesian coordinates. If $d(\mathbf{a}, \mathbf{b}) = \sqrt{\{(a_1 - b_1)^2 + (a_2 - b_2)^2 + (a_3 - b_3)^2\}}$ is the Euclidean metric for a pair of coordinate sets $\mathbf{a} = (a_1, a_2, a_3)$ and $\mathbf{b} = (b_1, b_2, b_3)$, then the quantity $d$ is the *distance* between the two particles whose coordinate sets are **a** and **b**, as measured by rigid, straight rods, or perhaps by lights rays, clocks and so on. Can anything more about the meaning of coordinates be said, however?

To get a deeper understanding of coordinates, it will be useful look briefly at the history of geometry. The early work on geometry, prior to 300 BC or so, was summarized by Euclid in the *Elements*. Euclid's work does not make use of coordinates, and indeed the fundamental concepts involved are those of point and (straight) line. It is assumed that a line segment, that is, the part of a line bounded by two points within it, can be translated in such a way as to make the relation of congruence an equivalence relation; thus some notion of rigidity is also assumed. Euclid's method is to lay down axioms concerning these basic notions of point and line, and thus describe the nature of physical space. One of the propositions that can be proved from the axioms is the famous theorem of Pythagoras, that the square on the hypotenuse of a right-angled triangle is equal to the sum of the squares on the other two sides.

Surprisingly perhaps, it was not until the seventeenth century AD that coordinate geometry was introduced, by Descartes (1637). Coordinates can be defined within the Euclidean framework as follows (in two dimensions, for simplicity). Take an arbitrary point $O$, the origin. Let the $x$-axis $Ox$ be an arbitrary line that includes $O$, and let the $y$-axis $Oy$ be the unique line that is perpendicular to $Ox$ and also includes $O$. Then the $x$-coordinate of a point $P$ is the perpendicular distance from $P$ to $Oy$, and the $y$-coordinate of $P$ is the perpendicular distance from $P$ to $Ox$. It may be shown from the axioms that any two points (in a plane) that share the same coordinate pair $(x, y)$ are coincident, so that the coordinates of a point determine its position uniquely.

The Euclidean metric, which gives the distance between two points as a function of their coordinate pairs, is also easily derived within Euclid's system. Suppose we wish to know the distance between $P$ and $Q$, where $P$ lies on $Ox$ and $Q$ is on $Oy$. Thus, the coordinates of $P$ and $Q$ are $(x_p, 0)$ and $(0, y_q)$ respectively. By the definition of coordinates, $x_p$ is the distance from $O$ to $P$, and $y_q$ is the distance from $O$ to $Q$. But, since $OPQ$ is a right-angled triangle, whose hypotenuse is $PQ$, the distance $PQ$ can be calculated using Pythagoras' theorem as $\sqrt{(x_p^2 + y_q^2)}$. This special case is easily strengthened to the general case, where $P$ and $Q$ may not lie on $Ox$ and $Oy$, by constructing ad hoc axes $Ox'$ and $Oy'$, parallel to $Ox$ and $Oy$, on which $P$ and $Q$ do lie. This yields the required result that the distance $PQ$ is equal to $\sqrt{\{(x_p - x_q)^2 + (y_p - y_q)^2\}}$, as $x'_p = x_p - x_q$, and $y'_p = y_p - y_q$.

Euclidean geometry thus includes the following proposition as a theorem, called CG (short for coordinate geometry).

**CG** There is a bijection between coordinate pairs $(x_p, y_p)$ and possible points $P$ such that the distance between any two points $P$ and $Q$ is equal to $\sqrt{\{(x_p - x_q)^2 + (y_p - y_q)^2\}}$.

It should be noted that CG does not explicitly say how coordinates are constructed, as perpendicular distances from coordinate axes. It gives no direct statement about what they actually represent.

A remarkable fact about CG is that it entails all the axioms (and hence all the theorems) of Euclidean geometry! The whole of Euclid's system is compressed into this one statement. CG even enables one to infer that $x_p$ is the perpendicular distance from $P$ to $Oy$, and so on – it is entirely self-sufficient.

CG involves only two empirical concepts, namely, point and distance. In view of the amazing brevity and simplicity of CG, it seems clear that point and distance are, in some sense, the fundamental concepts in geometry. CG sums up all the laws of geometry, which involve numerous concepts, into one pithy statement about points and distances.

As stated above, CG entails that individual coordinate numbers such as $x_p$ are perpendicular distances to coordinate axes, so we can give them a *direct* empirical significance, in addition to the indirect significance that the metric provides. It seems imaginable, however, that one could have coordinate sets with no direct empirical significance. The important thing is that, when pairs of them are fed into the metric, they generate the correct distances. It seems almost accidental that $x_p$, as well as $\sqrt{\{(x_p - x_q)^2 + (y_p - y_q)^2\}}$, is a distance. It is surely conceivable that the individual quantities $x_p$, $x_q$, and so on might have no interpretation outside of the metric.

If the numbers $x_p$, $x_q$, etc. had no direct empirical significance, then the manner of their correspondence to the concrete system would be rather mysterious. We would still be forced to posit such a correspondence, however, due to their ability to compress information about distances and predict the values of distances not yet measured.

### *7.3.4 The Nature of the State Vector*

In the case of state vectors, it does seem that the individual complex numbers have no direct empirical significance. The number $c_1$ in the vector $c_1|r_1\rangle + c_2|r_2\rangle$, for instance, is certainly not a chance as it is complex rather than real, and even its modulus is the square root of a chance rather than a chance. Moreover, although the phase of $c_1$ does not affect the chance $|c_1|^2$, it is still physically important in many situations, such as in the two-slit experiment, as it is needed to generate the correct chances. It seems that the only empirical significance of $c_1$ is its role in generating physical chances.

It will be useful to introduce the notion of a *substructure* here. In the case of geometry, we have seen that the axioms describing the properties of physical space can be summarized by the single proposition CG, which is a rule governing distances. The rule is formulated using another structure, however, the coordinate structure, from which distances can be derived. We might say therefore that the coordinate frame is a *substructure* for distance. In a similar way the rules of QM, which concern physical chance, are formulated in terms of wavefunctions, from which chances are derived. We might say therefore that the wavefunctions (or state vectors) are a substructure for chance.

One should not equate direct empirical significance with physical significance. Though state vectors seem to lack direct empirical significance, whereas coordinates have a direct empirical meaning, state vectors seem to be just as 'physical' as coordinates. As I argued in section 7.1.1, they seem to be descriptions of what is physically going on in a system. They prove their correspondence to the concrete system by generating the right values of the chance function, in a neat, economical, and compelling way. (Chance is, as shown in §4.6, an empirically accessible quantity.)

Suppose it is granted that a substructure for a physical quantity is also a physical description of a system. Thus, since distance is a physical quantity, we would accept coordinates as representative of the concrete system even if they had no direct empirical meaning. Similarly, since chance is a physical quantity, it follows that the wavefunction is a physical description. It should be remembered, however, that the physical chance function is (according to the causal theory of chance) an *epistemic* probability function. It is the probability function for an ideally rational agent who has maximal knowledge of certain physical causes. For this reason one might expect a substructure for chance to have some epistemological features.

We are faced with an interesting situation here, in that chance rather straddles the familiar division between physical facts and states of knowledge. It does represent a state of knowledge, but it is also fully determined by the physical facts. Chance is therefore both physical *and* epistemic. It is not somehow a mixture of the two, having a physical component and an epistemic component, but is *fully* physical and *fully* epistemic. How can this be? To understand this we must recall, from §2.2.2, that an objective state of affairs is defined in terms of epistemic states. The state of affairs $A$ is the expansion from $K_O$ to $K_A$. Thus, fundamentally, there is no distinction between physical facts and states of knowledge.[35]

One may object to the claim that *all* objective states of affairs are ultimately defined in terms of epistemic states on the grounds that it proves far too much. It is only in quantum mechanics that the states seem to have an epistemological character. In classical physics it quite a different story, as there no temptation to use terms like 'knowledge' and 'information' in the Copenhagen style. Why, on my account, does the epistemic nature of states in classical physics go unnoticed?

To see why classical states do not display any epistemic character, in spite of being fundamentally epistemic, we must review the argument, in §7.1.2, in favour of the epistemological interpretation of the $\psi$-function. The essential point was that the state vector is not localized in spacetime. In the case of the EPR experiment, for example, the two correlated systems are spacelike separated and yet neither has its own state vector. Since knowledge of two systems may well not be localized, it appears that the $\psi$-function represents knowledge about the two systems. In other words, the epistemological character of a state of affairs becomes apparent only when it does not *factorize*. When the state of a composite system factorizes into a set of states for its component subsystems, we simply do not notice anything epistemic going on. Now, as was argued in §6.3, the claim that states (and histories) of composite systems always factorize, namely CSM3, is one of the basic assumptions of classical (i.e., non-quantum) physics. Thus, in the classical domain, where this assumption holds FAPP, it is possible to ignore the epistemological character of the states of affairs.

In addition to the fact that states of affairs are themselves epistemological, we must remember that the chance function is an epistemic probability function. According to Definition 3.1.1, $P_X(A)$ is $Pr(A \mid \mathscr{L}_X \,\&\, bc_X)$. Thus, a chance is fundamentally a degree of belief, even though it is entirely determined by the physical facts $\mathscr{L}_X$ and $bc_X$.

This attempted *rapprochement* between physical states and states of knowledge may seem to threaten the very notion of objectivity, and the distinction between objective and subjective, but it is not so.[36] All I am claiming is that the concepts of knowledge and belief belong in both the objective and subjective realms. Human knowledge and belief are subjective, of course, but (as argued in §2.2) the logical distinction between Sinn and Bedeutung forces us to recognize the existence of objective counterparts of these human properties. It is only in this way that one can understand such objective, logical notions as entailment and truth. This idea of objective knowledge and belief does, of course, threaten some metaphysical views. One may well ask John Bell's question here: '*Information*? *Whose* information?' (Bell 1990: 34). We are not talking about the information (or knowledge) possessed by any human, so that my view seems to entail a minimal form of theism.

I have argued that the Hilbert space of state vectors is a substructure for physical chance, and that a substructure shares the general characteristics of the quantity for which it is a substructure. In this case, since chance is both physical and epistemic, it would follow that the wavefunction is also both physical and epistemological. This would explain why the wavefunction seems to have both physical and epistemological characteristics, as argued in §§7.1.1 and 7.1.2.

### *7.3.5 Time Evolution*

The example of intrinsic spins in section 7.3.2 ignores the time evolution of the state vector, during periods when **X** does not interact with any other system. In general, of course, the wavefunction evolves in time in accordance with the Schrödinger equation, which is linear and deterministic. How is this evolution to be understood, within the view that the wavefunction is a substructure for the chance function?

The Schrödinger equation is not only deterministic, but also has the property that a state vector always evolves in time to another state vector rather than something weaker, like a density matrix. This is quite striking, as it seems to mean that a saturated relative model of a system at time $t_1$ enables one to infer an equal amount of knowledge about the system at $t_2$ as a saturated model at $t_2$ would provide. The simplest way for this to happen would be for the saturated relative model itself to evolve deterministically (i.e., predictably) between external interactions. This is a little surprising, as it is common to think that determinism is a feature of the large-scale classical world, which QM forces us to renounce at the atomic scale. We often hear that individual atoms behave randomly, in unpredictable fashion, so that only certain statistics defined on large ensembles of them are predictable FAPP. This seems to be mistaken, however. The evolution of an atom's relative state is itself deterministic; it is only the outcome of its interactions with large systems that are unpredictable.

From our experience of geometry it is only to be expected that the evolution of quantum states will be modelled by variation of the state vector. The first precisely formulated dynamical laws, those due to Newton (1686), employed the geometrical coordinates of Descartes, modelling the motion of a particle by variation of its coordinates. While it is no doubt possible to formulate Newton's laws of motion and gravitation within a framework of synthetic geometry,[37] there is no doubt that the use of coordinates is a great convenience. In a similar way, it is only natural that the dynamical law of quantum mechanics is stated in terms of the variation of 'coordinates' with time.

## 7.4 SUMMARY

Where does this interpretation of the state vector, as a substructure for physical chance, leave us with regard to the problems described in section 7.1? These problems are

1 The tension between the physical and epistemic aspects of the state vector.
2 The AND/OR problem (i.e., the collapse of the wavefunction).
3 The 'shifty split' between system and apparatus.
4 The measurement problem (QM is metrocentric).

I do not claim to have provided complete solutions to all of these problems, but I think that my interpretation sheds some light on them.

Regarding (1), I argue that there is no tension between the physical and epistemological features of the state vector. The chance function, for which the state vector is a substructure, is both physical and epistemic in a way that is perfectly comprehensible. This does not help us to get much of an intuitive understanding of what the state vector means, of course. It gets rid of the tension rather than the mystery. It does not so much solve the mystery as describe it in precise terms.

The solution I offer to the AND/OR problem is not original in itself, although I hope that the material in §7.2, and particularly §7.2.4, helps to remove the usual objections to it. Suppose a small system is in a state represented as $|\psi\rangle = c_1|a_1\rangle + c_2|a_2\rangle$, where $|a_1\rangle$, $|a_2\rangle$ are normalized eigenstates of some operator $A$. Do the components $|a_1\rangle$ and $|a_2\rangle$ exist together in the concrete world, so that we should read the superposition as AND, or are they alternatives, only one of which really exists? In my view, neither of these options is exactly right. It is certainly wrong to treat the superposition as a disjunction, as if the system were more accurately represented by either $|\psi\rangle = |a_1\rangle$ + or $|\psi\rangle = |a_2\rangle$. One might as well consider a geometrical vector $a\mathbf{i} + b\mathbf{j}$ as a disjunction, so that only one of the components **i** and **j** is really 'there.' Both components are needed for a correct representation. When a measurement of $A$ is made, however, the chances of measurement outcomes are given by the squared moduli $|c_1|^2$ and $|c_2|^2$. Only one measurement outcome exists, of course, so at this stage we interpret the superposition as OR.

In short, this is the standard view that the state vector is reduced when a measurement occurs or, more generally, when the system becomes correlated in an appropriate way with a large system. This view has the consequence that either a large system does not have a state vector, or (if it does) that the Schrödinger equation is not universally valid. It seems to me that the former alternative is more likely, that systems large enough to have saturated models cannot be described by state vectors.

Since a measurement of **X** by **M** simply results in a correlation of **X** with **M**, why does this amount to an 'amplification to the classical level,' or 'definitization' of some quantity? Nothing of this sort takes place when **X** becomes correlated with another small system **Y**. The difference is that, since **M** has a saturated model *m*(**M**), the **X–M** correlation gives **X** a relative model *m*(**X/M**), which describes **X** alone. The correlation thus gives rise to a huge increase in the detail and completeness of **X**'s best model. The **X–M** correlation is like a lifeline connecting **X** to the macroscopic level.

The fact that, in QM, the act of measurement often causes a definite value for a physical quantity, a value that did not exist in the system prior to the measurement, strongly suggests that QM only applies to systems that are below the 'saturation level,' that is, to systems that do not have saturated models. These are systems whose best model is a relative model *m*(**X/M**) rather than an absolute model *m*(**X**). It is not clear to me even roughly where this saturation level may lie, however. Since it seems that, under many circumstances, decoherence induces FAPP classical behaviour in systems far below the classical level, this level is difficult to locate. Current research programs into macroscopic quantum tunnelling and coherence, where decoherence is suppressed by special conditions, may shed some light on this question, however.

If quantum mechanics only applies to small(ish) systems, describing them from the perspective of a large system they are correlated with, then there is clearly a need for a split between the system and the apparatus. The split is not an ontological one, of course, as the apparatus is also made of atoms, but rather a necessity of description. The apparatus is akin to a reference body, from whose perspective the system is described.

The shiftiness of the split is more a feature of the quantum-mechanical formalism than of reality. Suppose an apparatus **M** makes a complete measurement on a small system **X**. We can, of course, then treat ⟨**X**, **M**⟩ as a single system, and attempt to describe it from the perspective of another apparatus **N**, thus shifting the split. The two large systems **M** and **N** cannot be correlated, however, and **N** does not interact with **X**, and so it is not correlated with **X**. Thus, even if **N** makes a measurement of the measurement outcome of **M**, the best description of **M** is still just *m*(**M**), and the best model of **X** is still *m*(**X/M**). When **N** measures **M**, nothing of importance changes except the state of **N**.

If this situation is treated quantum-mechanically, under the assumption that the Schrödinger equation applies universally, then **N** ends up being correlated with ⟨**X**, **M**⟩. I maintain that this description is false, as no such chance correlation exists in fact. If the correlation did exist it would be very hard to observe, due to decoherence, so an empirical refutation of universal quantum mechanics may not be possible. In other words, decoherence allows one to put the split in the wrong place. Decoherence has a contrastive pair of roles, therefore. On the one hand, it makes quantum systems behave (under certain circumstances) in a classical way, and yet it also enables one to treat classical systems as if they obeyed quantum mechanics, without contradicting the empirical facts.

Finally, we come to the problem of measurement itself. Although I agree with Bell that QM applies to the world at large, and not just to 'piddling laboratory operations' (1990: 34), my general approach is to justify, rather than eliminate, the metrocentric nature of QM. Measuring devices, or more generally large systems, have an indispensable role in enabling the description of small systems. The best model of a small system is a model relative to a large system correlated with it.

The interpretation of the state vector proposed here is more complicated than most of its rivals, as it does not arise from just one idea. Although its foundation is the causal theory of chance, some additional components are required. The essential elements of this interpretation are as follows:

1 The causal theory of chance.
2 QM correlations are ultimately due to a *sui generis* relation of similarity between two systems. This means that their joint maximal description does not factorize, and so their joint utd chance function does not factorize either.
3 The fact that maximal descriptions need not factorize implies that there are logically emergent properties.
4 There exist systems with saturated models. Intuitively speaking, this is the level at which all properties have emerged already, that is, the 'classical level.'
5 The best model of a small system is a model relative to some large system (having a saturated model) correlated with it.
6 The wavefunction is a substructure for transition chances in a small system.

# 8

# Conclusion

This book covers a wide range of topics, so one may wonder what the overall message is. I think the main theme is that there cannot be a sharp separation between the study of probability by philosophers and the work of physicists. I have argued, in chapters 5 to 7, that certain problems of theoretical physics are greatly illuminated by a correct philosophical understanding of probability and its attendant notions. Moreover, though it may not be apparent in reading the book, the study of physical problems was of crucial importance in formulating the 'philosophical' analysis of chance in chapters 2 to 4. The chapters were not written, of course, in the order they appear above. In fact, the book really began with the fifth chapter, so that the analysis of chance was, to a large extent, made to fit a pre-existing system of mechanics. Indeed, the writing of the earlier chapters was influenced by the later chapters just as much as the reverse.

In the course of researching this book I have found it necessary to introduce some new notions, and resurrect some old ones. I shall end with a list of these, which I feel deserve recognition and further study.

1 We need a full-blooded, Aristotelian notion of efficient causation. The limp versions available within empiricism are not sufficient to do any physics.
2 In chapter 2 I sketched out a new approach to logic, based on epistemic states, inspired by the work of Peter Gärdenfors. I make use of this account to define logical probability, but I think that further study in this area would lead to many interesting discoveries. In particular, it seems that a workable correspondence theory of truth may now be possible.
3 The analysis of chance as logical probability brings the concept of epistemic state into physics, without falling into anthropocentrism. I believe that the consequences of this are very great indeed, and have tried, espe-

cially in chapters 6 and 7, to show the fertility of this idea in understanding quantum theory. It allows, for instance, non-factorizable states of a composite system, logically emergent properties, a distinction between predictive and causal locality, and relative states.

4 The concepts of lagrangian and boundary condition, familiar to physicists, need to be recognized as philosophically important as well. The notion of the generalized lagrangian enables one to define the determination relation in terms of entailment, and also to give an analysis of nomic necessity.

5 The distinction between abstract and concrete must be emphasized constantly. As noted in chapter 3, in the discussion of the symmetry-conservation argument, the very analysis of chance as partial determination depends on there being a noticeable difference between the concrete real world and the ersatz actual world. The idea of a perfect, abstract representation of a concrete system is a logical impossibility – there cannot be such a similarity between the real and the non-real. The premiss that even maximal states of affairs are incomplete is of vital importance in the work on quantum mechanics.

# Notes

## Chapter 1: Introduction

1 The idea of a partial cause is explained below, in §3.1.1. The rough idea is that, if two people together push a boulder off the top of a cliff, then each person is a partial cause of the boulder falling.
2 This term is used by Anscombe (1971: 25).
3 In order to avoid an infinite regress, one must hold that something is uncaused. This would be some rather special object, however, and not just a common physical event existing in spacetime.
4 The evidence for numerically precise chances comes from the stability of long-run relative frequencies, in exact repetitions of the same experiment. Thus, one cannot say for sure that chances are exact real numbers, as an experiment can only be performed a finite number of times. Also, only fairly simple experiments (involving simple systems such as photons) can be repeated exactly.
5 See Aspect et al. (1982).
6 See for instance Montague (1962) and Lewis (1979: 37).

## Chapter 2: Logic and Probability

1 This claim, notoriously, leads to the problem of the priors. In order to calculate $Pr(H \mid E)$ one has to use Bayes's theorem, for which one already needs the values of terms like $Pr(E \mid H)$ and $Pr(H)$. Unfortunately the terms like $Pr(H)$, the infamous prior probabilities, do not seem to exist.
2 This discussion is based on Howson and Urbach (1993: 59–72).
3 Rolf George (1997) identifies four distinct senses.
4 By 'sentence' I mean a declarative sentence, i.e., a sentence that has a truth value.

5 By 'state of affairs' I mean something like Reinach's *Sachverhalt*. For details, see Smith (1989).
6 Frege did not say this, but instead held that the meaning of a sentence is its truth value. I argue for my view over Frege's below.
7 Remarkably, these two explications of what a proposition is sometimes even appear together! See, for instance, Plantinga (1974: 45n1).
8 Though the subject of formal entailment is interesting, it does not concern us here.
9 Clearly, the statement that all epistemic states are deductively closed is part of the informal explication of epistemic states, and not a rigorous definition. The use of epistemic states to define entailment would otherwise be circular.
10 States of affairs are dependent on these ideal epistemic states in the same way that Gedanken are caused by human beliefs.
11 We could introduce a (fictitious) absurd epistemic state, to ensure that conjunctions always exist, so that the pure epistemic states form a Boolean algebra.
12 This notion of a tautology is not too different from that of Ellis (1976: 201), and may be objected to on the same grounds. Sorensen (1982) points out that, on this account, the proposition 'someone has a belief' is a tautology, which seems odd. Is the existence of a rational mind a *logical truth*? It seems to me that, in any system of logic founded upon epistemic states, the existence of a rational mind will be logically necessary. This is not to say that the existence of *human* minds is logically necessary, but it is likely that the epistemic logic developed here is inconsistent with naturalism.
13 Degrees of belief are defined in §2.3.
14 I actually got the idea from Plato, however, so its pedigree is not to bad. See the *Republic* (Cornford, 1941, VI 507–8).
15 One must be aware of the two meanings of 'measure.' First there is a specification of what it means for an object to be assigned a number. Second, there is the physical process by which a number is assigned to a particular object. I am using 'measure' in the first sense.
16 This is use of '<' does not fall within the scope of our previous definition, as $r$ is a quotient rather than a desirable object. One may imagine, however, that the symbol $r$ here denotes an arbitrary object of value $r$.
17 All values on this scheme are non-negative real numbers. The question of whether there are negative values does not concern us here.
18 Many logical truths cannot be known a priori.
19 This point was brought to my attention by Paul Bartha.
20 The first three of these axioms were formulated by Kolmogorov (1933).
21 The assumption of irrationality becomes dubious in the case of bets made at different times. It means that changing one's mind, without additional information, is always irrational. In a sense this is true, but frequently such changes are corrections

of earlier mistakes. Here, it is the original belief, rather than the change, which is irrational, yet it is the change (not the original belief) that exposes one to a Dutch book.

22 Since $A$ is a set of points, $fA$ is the set of images of these points under $f$, i.e., $fA = \{(x, y): f^{-1}(x, y) \in A\}$.

23 I am tempted to miss out the number 5 and call this Axiom 6, as 5 is an unlucky number for axioms. Consider the trouble with Frege's Rule V, Euclid's fifth postulate, and Peano's fifth axiom.

24 van Fraassen (1989: 292).

25 Howson and Urbach (1993: 71–2).

26 Carnap (1950).

27 Howson and Urbach (1993: 52). I have adjusted thc notation to harmonize it with the rest of this chapter.

28 It should be noted that $Pr(A \mid U)$ and $Pr(B \mid U)$ do not exist, as [\$1 if $A$] and [\$1 if $B$] do not have precise values. For instance, [\$1 if $A$] is clearly preferable to 10¢, and less preferable than 50¢, but it is impossible to say more than this.

29 This criticism of the classical Principle of Indifference is given in Keynes (1921: 42).

30 See Bertrand (1889).

31 Keynes (1921: 55–6); italics in original.

32 His reference (p. 54) to the case where the balls are drawn by a magnet, and the balls are made of different metals, suggests strongly that it is.

33 See Howson and Urbach (1993: 61–2).

**Chapter 3: Causation and Determination**

1 The content of this section has been greatly influenced by the work of Christopher Hitchcock, by means of a talk given at the University of British Columbia in the Fall of 2000. Hitchcock and I differ significantly on this topic, however.

2 Book II, available in Barnes (1984).

3 Strictly speaking, there will be small forces from air currents and so on, but these are small and we shall overlook them.

4 Of course it is likely that I am connected to those plants by *some* causal processes, constituted by sound waves, light waves, and so on, but these are irrelevant to the example. It is not by means of those processes that I am supposed to have killed the plants.

5 To keep things simple, I am neglecting consideration of moments. We should really consider the line of application of each force to ensure that E is also in moment equilibrium.

6 See Locke (1689), Book II, chap. XXIII, §2.

7 For simplicity, I am making the Newtonian assumption that spacetime is uniquely

decomposable into space and time. This assumption is not necessary for the view itself.

8 This view of time does not require there to be a final time slice, but there has to be an initial one. Every physical system needs a boundary condition in order to produce a real, concrete history. This boundary condition is the ontological foundation of the concrete history, and is automatically its initial state.

9 I can make no sense of the idea that there is something called the present that 'moves.' I take the 'block universe' view that terms like 'now,' 'present,' and so on are token-reflexive indexicals, like 'here.'

10 See Ramsey (1931a).

11 *A*(*K*) is the expansion of *K* by *A*. See §2.2.2 for the full meaning of this notation.

12 Note that, for such a *K* that contains only objective physical facts, the counterfactual conditionals have objective truth values.

13 We will always hold the dynamical laws fixed.

14 I believe it was first given by Montague (1962).

15 Note that the expression 'state of affairs concerning C' means one that concerns C only, and no other object.

16 Equivalently, *m*(C) is the conjunction of all the states of affairs concerning C.

17 This material is found in Book II, chap. 1 (192b–193b), and is available in Barnes (1984).

18 See, for example, Sposito (1976: chap. 9).

19 Note that, from the point of view of CTC, this inference is an application of Miller's Principle. A law is necessitated by the generalized lagrangian, so that its chance is one.

20 Or every *contingent* event, to avoid a regress.

21 Even quantum-mechanical systems are not *indisputably* stochastic, but the evidence for their being stochastic is very strong indeed.

22 The basic idea of this argument is described by van Fraassen (1989: 239–40).

23 The 'system' here includes the donkey, the bales, and all other objects that are causally relevant, in the sense of being able to interact with the donkey.

24 This sort of thing happens all the time in chemistry labs, where the mixing of two colourless liquids produces a red liquid, the mixing of two stable compounds produces a sudden explosion, and so on.

25 There may be, for instance, an arrow outside the donkey system that points at one of the bales. However this arrow, being outside the system, cannot interact with anything in the system (by hypothesis).

26 I am therefore in agreement with Hume on this point, at least. We cannot understand the ultimate nature of causal connections. We do nevertheless have *some* idea of efficient causation, which is sufficient to distinguish it from any logical relation, or from mere constant conjunction or habitual inference.

**Chapter 4: Physical Chance**

1 See, for instance, Lewis (1980).

2 Unfortunately there is no sensible name for this principle. 'The Principal Principle' is obviously silly, and van Fraassen's term, 'Miller's Principle' is also inappropriate because David Miller rejected the principle. I use the latter term simply because it is less of a mouthful.

3 See §3.2.2 for an account of the generalized lagrangian.

4 For the sake of brevity, I will henceforth use the term 'cause' in place of 'efficient cause.'

5 I am grateful to Paul Bartha for pointing out this aspect of the causal theory.

6 A time slice of a system **X** is, roughly speaking, the part of **X** that exists in some time interval $[t_1, t_2]$.

7 Of course chances themselves would still be contingent, as they depend upon the history of the world as well as on the theory of chance.

8 Note that the history-to-chance conditionals are not exactly of the form 'if the initial segment of the actual history is $H_t$, then the chance of $E$ is $p$.' Rather, they are like 'given that the initial segment of the actual history is $H_t$ (and the dynamics) one can infer $E$ to degree $p$.' The difference is not too important, however.

9 I suspect that he would not, although from the point of view of an applied mathematician this seems untenable.

10 Lewis, for instance, offers no criterion to pick out 'admissible' propositions, i.e., those propositions such that knowledge of them does not render Miller's Principle inapplicable.

11 I say 'appears,' since convergence is a property only infinite sequences can have, and furthermore is a 'tail event,' depending only on the final (infinite) segment. Breiman (1968: 40) defines a tail event as follows. 'Let $X_1, X_2, \ldots$ be any process. A set $E \in \sigma(\mathbf{X})$ will be called a tail event if $E \in \sigma(X_n, X_{n+1}, \ldots)$, for all $n$.'

12 I shall use 'frequency' as an abbreviation for 'relative frequency.'

13 The physics involved in this assumption is examined in detail in §5.4.

14 Of course one also needs a sufficiently uniform prior epistemic probability distribution over the possible values of the chance, but this is a different problem.

15 Thus, for instance, Lewis (1980: 106–8) has the resources to carry out this calculation, even though he does not endorse any particular theory of chance.

16 Of course one must also assume independence, so we really have $P_K(f(H) = 0.51 \mid P(H) = \frac{1}{2}$ & independence).

17 See E. Nagel (1961: 21).

18 I do not say 'and only if,' as there may be other kinds of explanation.

19 For the sake of the example I am assuming that the coin is genuinely stochastic, and

that the chance of heads is the same for each toss (i.e., independent of the boundary condition).

20 This is not strictly true in the case of an infinite sequence of experiments, but I shall not pursue the matter here.

21 This formulation is actually due to Alonzo Church (1940). It sharpens up von Mises's original idea.

22 I believe this acronym was first used by Bell (1990: 33).

23 See note 11.

24 For a detailed discussion of this objection see Howson and Urbach (1993: 331–7).

25 Howson and Urbach claim to have solved this problem, however. We will discuss their 'Bayesian reconstruction' of von Mises in §4.7.3 below.

26 More precisely, I mean its limiting relative frequency.

27 At least, this is how I read Popper. Van Fraassen somehow interprets Popper as holding that there is a *single* virtual sequence that would arise in an infinite sequence of trials. This interpretation makes Popper inconsistent of course, as he is committed also to indeterminism.

28 For simplicity, I am assuming that time is discrete.

29 See for instance McCurdy (1996).

30 This point is argued in §2.6.

**Chapter 5: Classical Stochastic Mechanics**

1 In this chapter we are only concerned with chance distributions for random vectors and variables, so there is no need to specify a sample space. The definition of a random variable as a $P$-measurable function $\Omega \rightarrow S$ is just a mathematical nicety; terms like $P(\mathbf{X} = \boldsymbol{x})$ are unambiguous even if $\Omega$ is unspecified. (Think of $\Omega$ as the class of all possible worlds, if you will.)

2 Note that I assume a 'B-series,' or 'block universe' view of time, where each time is the present from its own point of view. (See McTaggart 1908.) I can make no sense of the idea that time flows, or that the present moves, or that the present has a special metaphysical status.

3 See Snyder (1975: 186–7).

4 For simplicity, I am assuming that the law function assigns non-zero measure even to single histories, although in general we must of course deal with Borel classes of histories.

5 See Gardiner (1983: 117–74).

6 See Breiman (1968: 393).

7 $K_O$ is the state of maximal ignorance, defined in §2.2.2.

8 For details of $R$, the relative probability function, see §2.5.

9 Schrödinger (1935: 161) uses the term *Verschränkung unseres Wissens* to describe this situation.

10 Recall that, in §4.1, I argued that (at least some) boundary conditions represent *physical* constraints.
11 Sklar (1986), Lebowitz (1993, 36), and Feynman (1967, chap. 5).
12 Or something similar, such as a surface of space-like separated points.
13 For a fuller discussion of this problem, see Price (1996: 16–21), Sklar (1986), or Davies (1974).
14 This is in order to make the temporal direction a global, rather than merely local, one.
15 This matter is discussed in Arntzenius (1995).
16 This quantity is not exactly the Boltzmann entropy, of course, but it is related to it.
17 Note that, since on this model the particles' momenta are ignored, each state is its own time reverse.
18 We could always let **W** represent all these processes lumped together, if there were more than one.

**Chapter 6: Correlation**

1 This presentation of QM is partially based upon that of William Unruh (1994).
2 That is, a density operator that is not also a projector. Clearly, a projector is a special case of a density operator.
3 For relative phases of 0, $2\pi$, etc. the inference is constructive, but for a relative phase of $\pm\pi$ it is destructive.
4 Recall that $\langle A \rangle = Tr(\rho A)$.
5 This presentation is based on that of Leggett (1986: 35–6).
6 Landau and Lifshitz (1977: 2) put this point as follows: 'The possibility of a quantitative description of the motion of an electron requires the presence also of physical objects which obey classical mechanics to a sufficient degree of accuracy.'
7 This argument was first made by Heisenberg (1927: 68).
8 Bohr first met Einstein in 1920, and by Bohr's account (1949) their basic differences of approach were already apparent. The famous 'Bohr–Einstein dialogue' is usually considered to have begun in 1927, however, at the Fifth Physical Conference of the Solvay Institute.
9 For a discussion of Einstein's views on QM, see Ballentine (1970). There is, however, some controversy about Einstein's attitude to the statistical interpretation.
10 The proof given here differs from Bell's, and follows instead Mermin (1985).
11 This term 'local realism' is due to Bernard d'Espagnat (1979: 158). It is also defined here in §6.3.3.
12 See Mermin (1993: 813).
13 We gain some insight into the non-locality of Bohm theory here. A force on $X$ alters the momentum of $Y$, it seems, because $Y$'s momentum *resides* at $X$ just as much as it

resides at *Y*, if *X* and *Y* are correlated. It becomes difficult even to regard *X* and *Y* as distinct entities. Perhaps it is more natural to think of a single particle that moves through the configuration space. The spatial separation of *X* and *Y* then merely gives an illusion of ontological separation.

14 See, for instance, London and Bauer (1939, §11).

15 According to Albert (1994: 66), 'Many researchers have perennially dismissed Bohm's theory on the grounds that it granted a privileged mathematical role to the positions of the particles.'

16 For one approach to a hidden-variables quantum field theory see Bell (1984).

17 The picture is actually a little more complicated than this, as one must consider the issue of the memory of the observer, and matters of consistency between a second observation of the instrument reading and one's memory of the first observation, and so on. I refer the reader to Everett (1957) for the full details.

18 This is probably a virtue of Everett's approach, as macroscopic quantum phenomena are now being discovered. For an introduction to these developments see Leggett (1986). For more empirical detail see Stamp (1991, 1995, 1996).

19 For a discussion of actuality as a relation between times in Everett's theory, see Saunders (1995).

20 If we looked at the cardinal number, then 'most' worlds would have a relative frequency close to $\frac{1}{2}$, for well-known combinatorial reasons.

21 This example was given in a lecture course by William Unruh.

22 This view is advanced by Bell himself, for example (1964, 1975, 1980).

## Chapter 7: The State Vector

1 Schrödinger developed his equation long after Balmer's formula was published, but such calculations are still called 'predictions' by philosophers of science.

2 Of course coordinates, unlike distances, are only defined within a reference frame, so that distances are more absolute, but coordinates are nonetheless deeper.

3 Since the particle has only one property, namely its position, the configuration space for the system is just ordinary physical space. The plane wave is, of course, a momentum eigenstate in which the position of the particle is highly uncertain.

4 The term 'statistical interpretation' is used by Ballentine (1970), who claims that it is 'rather like [the opinion] of Einstein' (358).

5 Bohr actually uses the word 'complete,' but it seems to me that he really means to say that it is maximal.

6 The statistical interpretation is consistent with, but not essentially committed to, a deterministic evolution of the underlying micro-state. Thus, if even the complete state evolves stochastically, some transition chances will remain even when all ignorance probabilities are removed.

7 Ironically Schrödinger, though an opponent of the epistemological viewpoint, was perhaps its clearest expositor. See §§6–10 of Schrödinger (1935).

8 As noted in §6.2.3, Everett denies this claim, as do those who hold that conscious minds are required for such definite values to emerge.

9 See Gottfried (1991: 37).

10 For All Practical Purposes.

11 Note that, unless the measurement is 'ideal,' the final state of the measured system need not be an eigenstate of the observable just determined.

12 As Joos and Zeh (1985: 224) put it, 'The interference terms still exist, but they are not *there*!' (i.e., they are not localized in the system wavefunction).

13 From the Greek μετρον, meaning measure. It is intended to be parallel to such fun-to-use terms as 'Eurocentric,' 'logocentric,' and 'phallocentric.'

14 Bell hopes for a 'more objective description of nature.' This may involve a continued role for something like the wavefunction, but Bell thinks it much more likely that 'the new way of seeing things will involve an imaginative leap that will astonish us' (1966: 27).

15 See, for instance, Bell (1976). The GRW approach is sympathetically discussed in Bell (1987).

16 See Bohr (1958: 73).

17 The term 'amplification' is misleading, as it suggests that the measurement result already existed in the system prior to the measurement process, but was too small to see. This is not the case in general, however, when the system is not in an eigenstate of the observable being measured. What does not exist cannot be amplified.

18 Again, Everett denies this, as do those who hold that conscious observers collapse the wavefunction.

19 I am thinking of 'Schrödinger's Cat'-type situations, where an atomic-level superposition is amplified to the macroscopic level. This actual state is rather more complicated than a pure superposition, however, as the cricket ball rapidly becomes correlated with its environment, and so has to be described by a density matrix. The essential point is that the ball's velocity is not well defined.

20 The other words on the black list are *system*, *apparatus*, *environment*, *reversible*, *irreversible*, *observable*, *information* and, worst of all, *measurement*.

21 It is widely felt that size, in this sense, has something to do with complexity as well.

22 In a typical confusion of causation with determination, T. Nagel says, 'True causes *do* necessitate their effects: they make them happen or make them the case' (1979: 186). We have seen that making something happen is extensionally distinct from necessitating it, yet Nagel actually sees the two as *synonymous*!

23 Note that the maximal description of a system must include any associated fields, in addition to the material bodies. For instance, a body containing electrons generates an EM field that should be seen as part of the system.

24 This notation is not ideal, as it does not express the fact that the entailment fails in the opposite direction. When the entailment does hold both ways, however, I will use the symbol '⇔.'

25 We must be careful here, however, since some of this impression of value-definiteness may be due to decoherence in the system's density matrix.

26 If **X** were correlated with both **M** and **N**, then an even more complete description of **X** would be provided by the conjunction *m*(**X/M**) & *m*(**X/N**).

27 This suggestion is no doubt far too late, but the theories of special and general relativity should be renamed, perhaps as Einstein's theories of spacetime and gravitation. Similarly, instead of talking about a 'relativistic' theory, we should speak only of a Lorentz-invariant theory.

28 This relativity logic should not therefore be compared to Putnam's (1969) version of 'quantum logic,' which involves an ad hoc tinkering with some laws of classical logic, such as the law of distribution of 'and' over 'or.'

29 This statement is precise only in the case where the quantity measured is a constant of the motion. In general, the outcomes will differ slightly in the manner dictated by the Schrödinger equation.

30 See, for instance, Schrödinger (1935), §8.

31 Before the second measurement occurs, **X** is already correlated with **M** as a result of the first measurement.

32 I am thinking of the kind of seal that was once used to authorize official documents.

33 The conditioning on $m_t$(**X/M**) screens off the boundary condition, so the latter can be ignored.

34 This is most easily done using the Pauli spin matrices, which define the spin operators along the *x*, *y*, and *z* axes. Other spin matrices, for different axes, can be obtained as linear combinations of these three, using simple trigonometry. The two spinors for a given measurement apparatus are then just the normalized eigenvectors for the appropriate spin matrix. The eigenvector for spin up is the one with corresponding eigenvalue $\hbar$, and so on.

35 It must be remembered here that these are not *human* states of knowledge, but those of an ultimate rational principle, something like Plato's Form of the Good. Note also that Bohr often speaks of 'the knowledge' of a physical quantity rather than 'our' knowledge.

36 It is interesting that some physicists have held that QM threatens the very notion of objectivity. London and Bauer (1939: 220) describe this view as follows: '[T]he discussion of this formalism taught us that the apparent philosophical point of departure of the theory, the idea of an observable world, totally independent of the observer, was a vacuous idea.' I strongly disagree with this.

37 For a heroic attempt at such a formulation, see Field (1980).

# Bibliography


Aharonov, Y., P.G. Bergmann, and J.L. Lebowitz. 1964. 'Time symmetry in the quantum process of measurement.' *Physical Review* 134B, 1410–16.

Albert, D. 1994. 'Bohm's alternative to quantum mechanics.' *Scientific American*, May, 58–67.

Anscombe, G.E.M. 1971. *Causality and Determination*. Cambridge: Cambridge University Press.

Arntzenius, F. 1990. 'Physics and common causes.' *Synthese* 82, 77–96.

– 1993. 'The common cause principle.' *PSA Proceedings* 2, 227–37.

– 1995. 'Indeterminism and the direction of time.' *Topoi* 14, 67–81.

Aspect, A., J. Dalibard, and G. Roger. 1982. 'Experimental test of Bell's inequalities using time-varying analyzers.' *Physical Review Letters* 49, 1804–7.

Ballentine, E. 1970. 'The statistical interpretation of quantum mechanics.' *Review of Modern Physics* 42, 358–81.

Barnes, J., ed. 1984. *Complete Works of Aristotle*. Princeton: Princeton University Press.

Bealer, G. 1998. 'Propositions.' *Mind* 107, 1–32.

Bell, J.S. 1964. 'On the Einstein-Podolsky-Rosen paradox.' *Physics* 1, 195–200. Reprinted in Bell (1987), 14–21.

– 1966. 'The moral aspect of quantum mechanics.' In Bell (1987), 22–8.

– 'Subject and object.' In Bell (1987), 40–4.

– 1975. 'The theory of local beables.' In Bell (1987), 52–62.

– 1976. 'Einstein-Podolsky-Rosen experiments.' In Bell (1987), 81–92.

– 1980. 'Atomic-cascade photons and quantum-mechanical nonlocality.' *Comments on Atomic and Molecular Physics* 9, 121–6. Reprinted in Bell (1987), 105–10.

– 1982. 'On the impossible pilot wave.' *Foundations of Physics* 12, 989–99. Reprinted in Bell (1987), 159–68.

– 1984. 'Beables for quantum field theory.' In Bell (1987), 173–80.

– 1987. *Speakable and Unspeakable in Quantum Mechanics*. Cambridge: Cambridge University Press.

– 1990. 'Against "measurement."' *Physics World*, August, 33–40.

Bell, J.S. and M. Nauenberg. 1966. 'The moral aspect of quantum mechanics.' In Bell (1987), 22–8.

Bertrand, J. 1889. *Calcul des probabilités*. Paris: Gauthier-Villars.

Bohm, D. 1951. *Quantum Theory*. Englewood Cliffs, NJ: Prentice-Hall.

– 1952. 'A suggested interpretation of the quantum theory in terms of "hidden" variables, I and II.' *Physical Review* 85, 166–93. Reprinted in Wheeler and Zurek (1983), 369–96.

Bohr, N. 1935. 'Can quantum-mechanical description of reality be considered complete?' *Physical Review* 48, 696–702. Reprinted in Wheeler and Zurek (1983), 145–51.

– 1949. 'Discussion with Einstein on epistemological problems in atomic physics.' In *Albert Einstein, Philosopher Scientist*, 200–41. La Salle, Ill.: Open Court. Reprinted in Wheeler and Zurek (1983), 9–49.

– 1958. *Atomic Physics and Human Knowledge*. New York: Wiley.

Breiman, L. 1968. *Probability*. Reading, Mass.: Addison-Wesley.

Carnap, R. 1950. *Logical Foundations of Probability*. Chicago: University of Chicago Press.

Cartwright, N. 1979. 'Causal laws and effective strategies.' *Noûs* 13, 419–37.

Church, A. 1940. 'On the concept of a random sequence.' *Bulletin of the American Mathematical Society* 46, 130–35.

Cocke, W.J. 1967. 'Statistical time symmetry and two-time boundary conditions in physics and cosmology.' *Physical Review* 160, no. 5, 1165–70.

Cohen, R.S., and M.W. Wartofsky, eds. 1969. *Boston Studies in the Philosophy of Science*, vol. 5. Dordrecht: Reidel.

Cornford, F.M., ed. 1941. *The Republic of Plato*. Oxford: Clarendon Press.

Davies, P.C.W. 1974. *The Physics of Time Asymmetry*. London: Surrey University Press.

De Broglie, L. 1924. *Thèse*. Paris: Masson et Cie.

De Finetti, B. 1937. 'La prévision: Ses lois logiques, ses sources subjectives.' *Annales de l'Institut Henri Poincaré* 7, 1–68. English translation, 'Foresight: Its logical laws, its subjective sources,' in Kyberg and Smokler (1964).

Descartes, R. 1637. *Géométrie*. Indianapolis: Bobbs-Merrill, 1965.

D'Espagnat, B. 1979. 'The quantum theory and reality.' *Scientific American* 241, November, 158–81.

Earman, J. 1992. *Bayes or Bust*? Cambridge, Mass.: MIT Press.

Einstein, A. 1922. *The Meaning of Relativity*. London: Chapman & Hall, 1967.

Einstein, A., B. Podolsky, and R. Rosen. 1935. 'Can quantum-mechanical description of reality be considered complete?' *Physical Review* 47, 777–80. Reprinted in Wheeler and Zurek (1983), 138–43.

Ellis, B. 1976. 'Epistemic foundations of logic.' *Journal of Philosophical Logic* 5, 187–204.

Everett, H. 1957. '"Relative state" formulation of quantum mechanics.' *Reviews of Modern Physics* 29, 454–62. Reprinted in Wheeler and Zurek (1983), 315–23.

Feynman, R. 1967. *The Character of Physical Law*. Cambridge, Mass.: MIT Press.

Field, H. 1980. *Science without Numbers*. Oxford: Basil Blackwell.

Frege, G. 1884. *Die Grundlagen der Arithmetik*. English translation, *The Foundations of Arithmetic*. Oxford: Blackwell, 1950.

– 1891. 'Über Function und Begriff.' In Geach and Black (1980), 21–41.

– 1892a. 'Über Begriff und Gegenstand.' In Geach and Black (1980), 42–55.

– 1892b. 'Über Sinn und Bedeutung.' In Geach and Black (1980), 56–78.

Gärdenfors, P. 1988. *Knowledge in Flux*. Cambridge, Mass.: MIT Press.

Gardiner, C.W. 1983. *Handbook of Stochastic Methods for Physics, Chemistry, and the Natural Sciences*. New York: Springer-Verlag.

Geach, P., and M. Black. 1980. *Translations from the Philosophical Writings of Gottlob Frege*, 3rd ed. Oxford: Blackwell.

George, R. 1997. 'Psychologism in logic: Bacon to Bolzano.' *Philosophy and Rhetoric* 30, no. 3, 213–42.

Ghirardi, G.C., A. Rimini, and T. Weber. 1986. 'Unified dynamics for microscopic and macroscopic systems.' *Physical Review D* 34, no. 2, 470–91.

Gillies, D. 1973. *An Objective Theory of Probability*. London: Methuen.

Gottfried, K. 1991. 'Does quantum mechanics carry the seeds of its own destruction?' *Physics World*, October, 34–40.

Green, H.S. 1958. 'Observation in quantum mechanics.' *Nuovo Cimento* 9, no. 5, September.

Hacking, I. 1975. *The Emergence of Probability*. Cambridge: Cambridge University Press.

Halmos, P.R. 1950. *Measure Theory*. New York: Van Nostrand Reinhold.

Heisenberg, W. 1927. 'The physical content of quantum kinematics and mechanics.' In Zurek and Wheeler (1983), 62–86.

Hempel, C. 1965. *Aspects of Scientific Explanation*. New York: Free Press.

Horwich, P. 1987. *Asymmetries in Time*. Cambridge, Mass.: MIT Press.

Howson, C., and P. Urbach. 1993. *Scientific Reasoning: The Bayesian Approach*. 2nd ed. La Salle, Ill.: Open Court.

Hughes, R.I.G. 1989. *The Structure and Interpretation of Quantum Mechanics*. Cambridge, Mass.: Harvard University Press.

Hume, D. 1739. *A Treatise of Human Nature*. Oxford: Clarendon Press, 1967.

– 1748. *An Enquiry concerning Human Understanding*. Indianapolis: Hackett, 1977.

Hutchison, K. 1993. 'Is classical mechanics really time-reversible and deterministic?' *British Journal for the Philosophy of Science* 44, 307–23.

Jarrett, J.P. 1984. 'On the physical significance of the locality conditions in the Bell arguments.' *Noûs* 18, 569–89.

Jeffrey, R.C., ed. 1980. *Studies in Inductive Logic and Probability.* Vol. 2. Berkeley: University of California Press.

– 1983. *The Logic of Decision.* 2nd ed. New York: McGraw-Hill.

Joos, E., and H.D. Zeh. 1985. 'The emergence of classical properties through interaction with the environment.' *Zeitschrift für Physik B – Condensed Matter* 59, 223–43.

Jordan, P. 1949. 'On the process of measurement in quantum mechanics.' *Philosophy of Science* 16, 269–78.

Keynes, J.M. 1921. *A Treatise on Probability.* London: Macmillan.

Kolmogorov, A.N. 1933. *Foundations of the Theory of Probability.* New York: Chelsea, 1950.

Kyberg, H.E., Jr. 1974. *The Logical Foundations of Statistical Inference.* Dordrecht: D. Reidel.

Kyberg, H.E., Jr., and E. Smokler, eds. 1964. *Studies in Subjective Probability.* New York: John Wiley.

Landau, L.D., and E.M. Lifshitz. 1977. *Quantum Mechanics.* Oxford: Pergamon Press.

Lebowitz, J.L. 1993. 'Boltzmann's entropy and time's arrow.' *Physics Today,* September, 32–8.

Leggett, A. 1984. 'Schrödinger's cat and her laboratory cousins.' *Contemporary Physics* 25, no. 6, 583–98.

– 1986. 'Quantum mechanics at the macroscopic level.' In J. de Boer et al., eds, *The Lesson of Quantum Theory,* 35–57. UK: Elsevier Science Publishers B.V.

Levi, I. 1980. *The Enterprise of Knowledge.* Cambridge, Mass.: MIT Press.

Lewis, C.S. 1947. *Miracles.* 10th impression. Glasgow: Collins, 1985.

Lewis, D. 1973. 'Causation.' *Journal of Philosophy* 70, 556–67. Reprinted in Lewis (1986b), 159–71.

– 1979. 'Counterfactual dependence and time's arrow.' In Lewis (1986b), 32–51.

– 1980. 'A subjectivist's guide to objective chance.' In Lewis (1986b), 83–113.

– 1986a. *On the Plurality of Worlds.* Oxford: Blackwell.

– 1986b. *Philosophical Papers Volume II.* New York: Oxford University Press.

Locke, J. 1689. *An Essay concerning Human Understanding.* Oxford: Oxford University Press, 1975.

Lockwood, M. 1989. *Mind, Brain and the Quantum.* Oxford: Blackwell.

London, F., and E. Bauer. 1939. 'The theory of observation in quantum mechanics.' In Wheeler and Zurek (1983), 217–59.

McCulloch, G. 1989. *The Game of the Name.* Oxford: Clarendon Press.

McCurdy, C. 1996. 'Humphreys's paradox and the interpretation of inverse conditional propensities.' *Synthèse* 108, no. 1, 105–25.

McTaggart, J.M.E. 1908. 'The unreality of time.' *Mind* 18, 457–84.

Mellor, D.H. 1971. *The Matter of Chance*. Cambridge: Cambridge University Press.

– 1986. 'Fixed past, unfixed future.' In Barry Taylor, ed., *Contributions to Philosophy: Michael Dummett*. The Hague: Nijhoff, 1986.

Mermin, N.D. 1985. 'Is the moon there when nobody looks? Reality and the quantum theory.' *Physics Today*, April, 38–47.

– 1993. 'Hidden variables and the two theorems of John Bell.' *Reviews of Modern Physics* 65, no. 3, July, 803–15.

Mises, R. von 1928. *Probability, Statistics and Truth*. London: George Allen & Unwin, 1939.

Montague, R. 1962. 'Deterministic theories.' In *Decisions, Values and Groups*. Oxford: Pergamon Press.

Nagel, E. 1961. *The Structure of Science*. New York: Harcourt, Brace and World.

Nagel, T. 1979. *Mortal Questions*. Cambridge: Cambridge University Press.

Newton, I. 1686. *Principia Mathematica*. Berkeley: University of California Press, 1934.

Papineau, D. 1992. 'Can we reduce causes to probabilities?' *PSA Proceedings* 2, 238–52.

– 1997. 'Uncertain decisions and the many-minds interpretation of quantum mechanics.' *The Monist* 80, no. 1, 97–117.

Penrose, O., and I.C. Percival. 1962. 'The direction of time.' *Proceedings of the Physical Society* 79, 605–16.

Penrose, R. 1989. *The Emperor's New Mind*. Oxford: Oxford University Press.

– 1994. *Shadows of the Mind*. London: Vintage, 1995.

Plantinga, A. 1974. *The Nature of Necessity*. Oxford: Clarendon Press.

Polkinghorne, J.C. 1984. *The Quantum World*. London: Penguin.

Popper, K. 1934. *The Logic of Scientific Discovery*. London: Hutchinson, 1959.

Putnam, H. 1969. 'Is logic empirical?' In Cohen and Wartofsky (1969), 181–241.

Price, H. 1996. *Time's Arrow and Archimedes' Point*. New York: Oxford University Press.

Ramsey, F.P. 1931a. 'General propositions and causality.' In *The Foundations of Mathematics and Other Logical Essays*. London: Routledge and Kegan Paul.

– 1931b. 'Truth and probability.' In *The Foundations of Mathematics and Other Logical Essays*.

Redhead, M. 1987. *Incompleteness, Nonlocality and Realism*. Oxford: Clarendon Press.

Reichenbach, H. 1935. *The Theory of Probability*. Berkeley: University of California Press, 1949.

– 1963. *The Direction of Time*. Berkeley: University of California Press.
Rosenfeld, L. 1967. 'Niels Bohr in the thirties: Consolidation and extension of the conception of complementarity.' In S. Rozental, ed., *Niels Bohr: His Life and Work as Seen by His Friends and Colleagues*. Amsterdam: North-Holland, 1967.
Royden, H.L. 1963. *Real Analysis*. New York: Macmillan.
Salmon, W. 1984. *Scientific Explanation and the Causal Structure of the World*. Princeton, NJ: Princeton University Press.
Saunders, S. 1995. 'Time, quantum mechanics and decoherence.' *Synthèse* 102, 235–66.
Savage, L.J. 1954. *The Foundations of Statistics*. New York: John Wiley.
– 1962. 'Subjective probability and statistical practice.' In G.A. Barnard and D.R. Cox, eds, *The Foundations of Statistical Inference*, 9–35. New York: John Wiley.
Savitt, S. 1995. *Time's Arrow Today*. Cambridge: Cambridge University Press.
– 1996. 'The direction of time.' *British Journal for the Philosophy of Science*, 347–70.
Schrödinger, E. 1926. 'Quantization as an eigenvalue problem.' In G. Ludwig, ed., *Wave Mechanics*. Oxford: Pergamon, 1968.
– 1935. 'Die gegenwärtige Situation in der Quantenmechanik.' *Naturwissenschaften* 23, 807–12, 823–8, 844–9.
– 1951. *Science and Humanism: Physics in Our Time*. Cambridge: Cambridge University Press.
Sklar, L. 1986. 'The elusive object of desire: In pursuit of the kinetic equations and the second law.' *PSA Proceedings* 2, 209–55.
Smith, B. 1989. 'Logic and the *Sachverhalt*.' *The Monist*, 52–69.
Snyder, D.L. 1975. *Random Point Processes*. New York: John Wiley.
Sober, E. 1993. *Philosophy of Biology*. Boulder, Colo.: Westview Press.
Sorensen, R.A. 1982. 'Epistemic and classical validity.' *Journal of Philosophical Logic* 11, 459–60.
Stamp, P.C.E. 1991. 'Quantum dynamics and tunneling of domain walls in ferromagnetic insulators.' *Physical Review Letters* 66, no. 21, 2802–5.
– 1995. 'Time, decoherence, and "reversible" measurements.' In Savitt (1995), 107–54.
– 1996. 'Tunneling secrets extracted.' *Nature* 383, September, 125–7.
Stove, D.C. 1986. *The Rationality of Induction*. Oxford: Clarendon Press.
Suppes, P. 1970. *A Probabilistic Theory of Causality*. Amsterdam: North-Holland.
Tarski, A. 1935. 'On the concept of logical consequence.' In *Logic, Semantics, Metamathematics*. Oxford: Oxford University Press, 1956.
Unruh, W. 1990. 'Particles and fields.' In J. Audretsch, and V. de Sabbata, eds, *Quantum Mechanics in Curved Space-Time*, 89–110. New York: Plenum Press.
– 1994. Lecture notes to Physics 555B, unpublished.
Van Fraassen, B.C. 1980. *The Scientific Image*. Oxford: Oxford University Press.

– 1989. *Laws and Symmetry*. New York: Oxford University Press.
– 1991. *Quantum Mechanics: An Empiricist View*. Oxford: Clarendon Press.
von Neumann, J. 1932. *Mathematische Grundlagen der Quantenmechanik*. Berlin: Springer. Reprinted in English in Wheeler and Zurek (1983), 549–647.
Wheeler, J.A. 1981. 'Delayed-choice experiments and the Bohr-Einstein dialog.' In Wheeler and Zurek (1983), 182–200.
Wheeler, J.A., and H.Z. Zurek, eds. 1983. *Quantum Theory and Measurement*. Princeton, NJ: Princeton University Press.
Wigner, E.P. 1961. 'Remarks on the mind-body question.' In Wheeler and Zurek 1983), 168–81.
– 1963. 'The problem of measurement.' In Wheeler and Zurek (1983), 324–41.

# Index

Abstract objects. *See* Logical entities
Accidental generalization, 68
Admissible proposition, 90, 239n10
Albert, D., 242n15
Almost surely, 93
Amplification to the macroscopic level, 206, 231, 243n17
And/or problem, 196–200
Anscombe, G.E.M., 53, 72, 235n2
Anthropocentric, 10, 19, 85, 233
Anti-realism, 159–60
Aristotle, 7, 54, 58, 66, 79, 233
Arntzenius, F., 241n15
Arrow of time. *See* Time's arrow
Aspect (experiment) 4, 148, 169, 174, 235n5
Authority: argument from, 35; epistemic, 27; of logic, 10, 27, 52, 85, 89
Authority of chance. *See* Miller's Principle
Axioms of probability, 33–6

B-series view of time, 61, 134, 238n9, 240n2
Ballentine, L., 193, 241n9, 242n4
Balmer series, 150, 189, 242n1
Barometer example, 145
Bartha, P., 236n19, 239n5
Bayes's theorem, 94, 235n1
Beables, 205
Beacon, rotating, 55
Bedeutung, 13–18, 19, 26
Begriff, 18
Belief: context, 16; degree of, 9, 11, 29–33
Bell, J.S., 166, 204, 205, 216, 228, 231, 240n22, 242n22, 243n14; black list, 207, 243n20; inequality, 144, 148–9, 165–7, 169, 174, 178–9, 196
Bernoulli, J., 46
Bertrand's paradoxes, 47, 50–2
Betting quotient. *See* Contract, value of
Big Bang, 126
Bohm, D., 149, 166, 169–75, 195, 241n13, 242n15
Bohr, N., 150, 158–9, 160, 161, 168–9, 192–6, 205, 206, 214, 242n5, 244n35
Bohr-Einstein dialogue, 241n6
Boltzmann entropy, 241n16
Boolean algebra, 19, 236n11
Born, M., 150, 189; measure, 179–80
Boundary condition, 61, 67, 68, 86, 111–13, 123–32, 135; as physical

constraint, 86, 123–6, 135, 238n8
Bra vector space, 151
Brownian motion, 112
Bundle of goods, 30, 31
Buridan's ass, 74–5

Carathéodory technique, 113
Cardinal problem, 98, 99, 104, 106
Carnap, R., 11, 12, 45
Causal: interaction, 80, 118; mechanism, 73, 76; process, 55, 72, 73, 77, 134, 237n4; theory of chance, 86, 90, 109, 110, 232
Causation: by double prevention, 56; efficient, 54–5, 60, 63–4, 68, 70, 85, 164–5, 208–9, 233, 238n26; generic, 56–7; by omission, 56; and time, 61–2, 133–5
Church, A., 240n21
Classical physics, 81, 206
Classical world, 177–9
Coherent quantum state, 156
Collapse of wavefunction, 152, 181, 198, 200, 203, 204, 207, 217
Collective, 97, 101; deceitful vs. honest, 99–100, 102
Complementarity, 159
Complete, 81, 119, 209; vs. maximal, 5, 77–80, 83, 242n5; set of observables, 219; state of affairs, 77–80, 81, 83, 119, 209
Complex amplitudes. *See* Probability amplitudes
Composite systems, 115, 118–21, 153, 181, 183, 195
Concreteness. *See* Real existence.
Concrete objects, 58–61, 63–4, 65, 76, 79
Conditional probability, 35, 106–8, 131
Conditionals, 62–3, 238n12
Conditioning, principle of, 35, 36, 106
Configuration space, 150, 151, 170, 194, 242n13, 242n3
Conjunction, 23, 28, 81
Connected partition, 127, 129
Consciousness, 61, 177, 203, 243n8
Consistent, 23
Consistent family, 181–6
Constant conjunction, 238n26
Contract, 29, 31; value of, 29, 30
Contraction, 21
Convergence, axiom of, 97
Coordinates, 190, 222–6, 229; realism about, 190
Copenhagen interpretation, 158–60, 168, 176, 188, 192–6, 199–200, 228
Correlation: capacity, 220; in chance, 114, 148–87; classical chance, 149–50, 167, 181; epistemic, *see* Relevance; in frequency, 114–15
Cosmos, 61, 116, 125–6, 133
Counterfactual dependence, 56, 62–3
Cricket ball, 207, 243n19
Criterion of physical reality, 161, 162, 164, 168, 169, 187

Davies, P., 241n13
De Broglie, L., 150
Decoherence, 173, 177, 204, 231, 232, 244n25
Dedekind cuts, 32, 33, 37, 93
Deductive closure, 19
Delayed choice experiment, 196–8, 201, 206
Density operator, 154–6, 203, 244n25
Descartes, R., 225, 229
D'Espagnat, B., 186–7
Determinate, 82–3
Determination, 5, 64–9, 164–5, 208–9
Determined Effect Principle, 70

Determinism, 69–70, 123–6
*Deus ex machina*, 135
Diffusion, 137, 141–4
Disjunction, 22
Double-slit experiment. *See* Two-slit experiment
Dutch book, 33, 237n21
Dynamical: facility, 111, 113, 124, 129; laws, 62, 66–8; nature, 66–9, 86, 111; properties, 61, 89

Ehrenfest urn, 141
Eigenstates, 151
Einstein, A., 109, 160–1, 163, 164, 192, 241n9, 242n4
Element of reality, 161, 163, 167
Ellis, B., 21, 236n12
Emergence: causal vs. logical, 207–10; logical, 82, 207–10, 211, 232, 244
Empirical access, problem of, 99–100, 102–4, 105, 106
Empirically accessible, 91, 92, 94, 96
Energy methods, 66
Entailment, 5, 18–25, 65
Entanglement of knowledge, 118, 240n9
Entropy, 110, 125, 133, 141–4
Epistemic state, 21
Epistemological viewpoint. *See* Copenhagen interpretation
EPR: argument, 148, 160–4, 180; correlations, 175, 178–9; experiment, 8, 166–7, 169, 179, 187, 195
Equilibrium (thermal), 142–3
Equi-probability, 12, 45–9
Euclidean metric, 190, 222, 224–5
Event from nowhere, 4, 73, 74
Everett, H., 157, 161, 175–80, 201, 214, 243n8
Exact similarity, 76
Expansion, 21
Experimental arrangement, 158–9, 168
Explanation, 73, 76, 91, 95, 125, 175

Factorize, 82, 117–18, 120, 153, 181–6, 195, 206, 209, 232
FAPP. *See* For All Practical Purposes
Feynman, R., 72, 241n11
Field, H., 244n37
Final superposition, 173, 179
Flow of time, 134–5
Fokker-Planck equation, 113
Footprint example, 139
For All Practical Purposes (FAPP), 98–9, 173, 177, 181
Formal cause, 79
Frege, G., 13–16, 20, 236n6
Frequency theories of chance, 96–106
Fuzzy. *See* Determinate

Gambles. *See* Contract
Gärdenfors, P., 6, 20–2, 233; functions, 24, 28
Gardiner, C., 240n5
Gedanke, 14–16, 19, 59, 236n10
Generalized lagrangian, 66–9, 75, 77, 86, 95, 111, 123, 135–6
General relativity, 85, 109
Geometry, 190
George, R., 235n3
Ghirardi, G.C., A. Rimini, and T. Weber, 181, 199, 200, 243n15
God, 79, 228
God's-eye view, 159, 205, 215
Good family, 104
Goods, 29
Good sense, 13, 49
Gottfried, K., 243n9
Gravity, 66
Great Beyond, 126
Grundlagen der Arithmetik, 13, 14

Hamiltonian mechanics, 189, 194
Hamilton's Principle, 67
Heisenberg, W., 161, 192, 241n7
Heisenberg uncertainty principle, 153, 159, 160, 192
Hesperus/Phosphorus problem. *See* Identity, problem of
Hidden variables, 148–9
Hitchcock, C., 237n1
Hockey team example, 220
Howson, C., and P. Urbach, 45, 90–1, 101, 102–4, 235n2, 240n25
Howson's Principle, 102
Hume, D., 238n26
Hydrogen atom, 150

Ideal ensemble, 160, 192
Ideal thinker, 9, 20, 21, 24, 29, 65
Identity, problem of, 15–18
Incommensurable. *See* Preference
Independence: causal, 49, 51, 114; logical, 28, 49; principle, 31. *See also* Correlation
Indifference, Principle of, 10, 39, 45–8, 88, 237n29
Indivisible alternatives, 49
Inference problem, 68, 180
Inferior knowledge, 22
Infinite lottery, 37
Infinitesimal value, 32, 37
Intelligible object. *See* Logical entity
Interference: meddling, 137, 139; of waves, 155, 156, 191, 196–7
Interval probabilities, 37–8
Irreducible relation, 82
Isometry, 39, 41, 42

Joos, E., and H.D. Zeh, 204, 243n12
Jordan, P., 204

Ket vector, 151
Keynes, J.M., 11, 12, 13, 45, 48, 49
King Lear, 13
Kolmogorov, A., 236n20

Lagrange, J. 66
Lagrangian, 67. *See also* Generalized lagrangian
Landau, L., and Lifshitz, E., 241n6
Langevin's equation, 112
Laplace's rule of succession, 12
Large systems, 206–12
Law function, 110–14, 129, 141
Law of large numbers, 93
Learning terminus, 22
Lebowitz, J., 241n11
Leggett, A., 241n5, 242n18
Leibniz, G., 79
Length, 182–3
Lesbegue measure, 47
Levi, I., 90
Lewis, D., 64, 84, 88–90, 98, 235n6, 239n1, 239n15
Local realism, 167, 186–7
Locality: causal, 8, 161, 162, 164–5, 168, 169, 174, 176, 178, 195–6; of information, 210; predictive, 164–5, 166–8, 174, 176, 178, 186–7; of states of affairs, 80, 195
Loch Ness monster, 28
Locke, J., 61
Lockwood, M., 214
Logical entities, 58–61, 65, 76–7, 79, 234
Logical omniscience, 30
Logical probability, 5, 6, 9, 26, 68, 85–7
Logical space, 22
London, F., and Bauer, E., 242n14, 244n36
Lorentz invariance. *See* Relativity

Lunch box example, 218

Macroscopically local observables, 201, 202, 207
Macroscopic quantum phenomena, 177, 231, 242n18
Macrostate, 142
Manner of presentation, 15, 16
Many worlds interpretation, 214
Markov property, 113, 121–3, 127, 128, 131
Mass (analogy of), 85
Maximal: epistemic state, 22, 77–80; information, 192, 194, 200, 217; state of affairs, 20, 65, 67, 82, 83
Maze example, 87
McCurdy, C., 240n29
McTaggart, J., 240n2
Measure, 30, 236n15
Measurement, 172, 198–200, 201–4, 215–19; apparatus, 159, 205, 215–19; definite outcome of, 173; ideal, 216, 217, 243n11; problem, 204–6
Memory of the observer, 242n17
Mereology, 59, 81, 208–9
Mermin, D., 241n10
Meßwirklichkeit, 193–4, 205
Metrocentric, 205, 232
Miller's Principle, 6, 7, 9, 27, 84, 88, 91, 94, 124, 175, 180; proofs of, 89–90, 102–3, 238n19, 239n2
Minimal epistemic state, 24
Minkowsky, H., 215
Mixed quantum state, 154–6
Modal realism, 79
Montague, R., 235n6, 238n14
Mugs of coffee example, 182, 212

Nagel, E., 239n17
Nagel, T., 208–9, 243n22
Naturalism, 236n12
Natural motion. *See* Dynamical nature
Newton, I., 66, 67, 68, 71, 84, 91, 109, 160, 208, 229, 237n7
Nickel and dime example, 46
Non-linear dynamical systems, 74
Nonseparability, 176
Norm-squared rule, 173–5

Objectivity, 3, 18, 228, 244n36
Object of thought. *See* Logical entity
Ontological source, 58
Ontological support, 7, 58, 238n8
Opaque to the understanding, 64
Orthodox interpretation. *See* Copenhagen interpretation
Orthonormal basis, 154, 180, 201

Painter example, 57–8
Paradoxes of indifference, 45
Partial cause, 3, 54, 57, 235n1
Partial entailment. *See* Logical probability
Pauli spin matrix, 244n34
Penrose, R., 181, 200
Perfect. *See* Complete
Phase space, 150
Phase of a wave, 155
Physical interpretation of state vector, 189–91, 199
Plantinga, A., 236n7
Plato, 236n14, 244n35
Point-normalization, 128, 129
Polkinghorne, J., 168–9
Popper, K., 100–2, 109
Possible worlds, 16, 19, 20, 23
Preference: equivalent, 30; incommensurable, 30, 37, 46; strict, 30
Present (time), 61–2, 134, 238n9, 240n2
Price, H., 241n13

Principal principle. *See* Miller's Principle
Prior probability, 12–13, 45, 235n1, 239n14
Probabilistic theory of causation, 4, 5
Probability amplitudes, 189, 190, 194
Production. *See* Causation, efficient
Projection: operator, 154, 198; postulate, 152, 173, 176
Propensity, 100
Property, 59, 64, 75
Proposition, 15, 18–19
Psychologism, 13, 18
Pure epistemic state, 23
Pure quantum state. *See* Superposition
Putnam, H., 244n28
Pythagoras's theorem, 225

Qualitative identity. *See* Exact similarity
Quantum logic, 244n28
Quantum mechanics (QM), 72, 73, 77, 80, 83, 91, 96, 109; rules of, 150–8

Radioactive decay, 3, 72, 76, 107
Ramsey, F., 62
Randomness, axiom of, 97
Random walk, 87
Rational belief systems, 21
Real existence, 60, 64, 234
Reference: class, 101, 102; frame, 160, 205, 214
Reichenbach, H.: common cause principle, 7, 110, 137, 144–7; frequency theory of probability, 96
Reinach, A., 236n5
Relative: frequency, 91–6, 96–106, 114, 235n4; probability, 36–7; quantum state, 157–8, 161, 175–80, 203; state of affairs, 27–9, 212–15, 220, 231
Relativity, 174, 214, 244n27
Relevance, 114–18, 119
Revision, 21, 62
Rosenfeld, L., 168

Sachverhalt, 236n5
Salmon, W., 54, 55
Saturated model, 210–12, 220, 231
Saunders, S., 214, 242n19
Scalar product, 151
Schrödinger, E., 150, 193, 240n9, 242n1, 243n7; S.'s cat, 243n19; wave equation, 150, 169, 170, 176, 189, 200, 202, 229, 230, 231, 244n29
Scientific realism, 144
Scientific reasoning, 11–13
Screening off, 89
Sheet of paper, torn, 185, 212, 219
Shifty split, 200–4, 205, 206, 231
Single-case probability, 100, 101
Single out, 41, 42, 44
Sinn, 13–18, 26
Sklar, L., 241n11, 241n13
Small systems, 206–12, 212–15, 219–20
Smeared out, 194
Smith, B., 236n5
Snyder, D., 240n3
Solipsism, 178
Sorensen, R., 236n12
Spin-½ particles, 166, 223
Spontaneous event. *See* Event from nowhere
Spread of concreteness, 64
Stamp, P., 204, 242n18
State, 121–2
State of affairs, 14, 15, 59, 64, 236n10
State vector, 151, 226–9
Statistical interpretation of QM, 192, 196, 241n9, 242n4, 242n6
Statistical mechanics, 160, 179–80
Stern-Gerlach apparatus, 172

Stochastic equation of motion, 112–13
Subjective, 18
Substance, 61
Substructure, 227, 229
*Sui generis* relations, 185, 212
Superior knowledge, 22
Superposition, 152, 154, 155–6, 176, 198–200
Support, 57–8, 61
Symmetry: axiom, 11, 38–44; conservation, 5, 73–7, 83; geometrical, 39–43; logical, 43, 44, 48

Tail event, 99, 106, 239n11
Tarski, A., 20
Teleology, 57, 66
Theism, 228
Thought. *See* Gedanke
Time: arrow of, 125–6, 133, 135, 137–47; and causation, 61–2; reversal invariance, 136, 143, 144; slices, 61, 87, 123, 127, 131, 239n6; symmetry, 125–6, 133, 135–6, 139; translation invariance, 136
Total cause, 54–5, 57
Trace (of an operator) 154
Traces (of interference) 110, 133, 137
Transition probabilities, 110, 132, 133, 137–9, 142, 190, 222–4, 232
Truth, 26, 233; and belief, 18; set, 25
Two-slit experiment, 156, 191

Uncaused event. *See* Event from nowhere
Uncertainty (quantum), 152, 160
Unit, 30, 32, 183
Universal Causation, Principle of, 70, 74
Unruh, W., 241n1, 242n21
Up-to-date chance, 131, 152, 221
Urn sampling examples, 12, 38, 48–9, 105

Value: function, 31, 32; intrinsic, 31
van Fraassen, B., 45, 68, 96, 104–6, 144, 238n22, 240n27
Vector space notation, 151, 223
Vectors of knowledge, 22
Virtual image, 15
von Mises, R., 96, 97–100, 101, 109
von Neumann, J., 201–4; chain, 203

Wavefunction, 150
Waves of knowledge, 189, 191
Wax tablet, 218–19
Wheeler, J., 194, 196
Wigner, E., 194

Zeno's paradox of the arrow, 122